Physical Chemist. y
A Guided Inquiry

Thermodynamics

James N. Spencer
Franklin & Marshall College

Richard S. Moog
Franklin & Marshall College

John J. Farrell
Franklin & Marshall College

Houghton Mifflin Company   **Boston**   **New York**

Publisher: Charles Hartford
Executive Editor: Richard Stratton
Assistant Editor: Danielle Richardson
Editorial Assistant: Rosemary Mack
Senior Project Editor: Fred Burns
Manufacturing Manager: Florence Cadran
Marketing Manager: Katherine Greig

Printed in the U.S.A.

ISBN: 0-618-30853-9
1 2 3 4 5 6 7 8 9—DBH—07 06 05 04 03

To the Instructor

Physical Chemistry: A Guided Inquiry: Thermodynamics is not a textbook. This book is meant to be used in class as a *guided inquiry*. Much research has shown that more learning takes place when the student is actively engaged, and ideas and concepts are developed by the student, rather than being presented by an *authority* — a textbook or an instructor[1]. The *ChemActivities* presented here are structured so that information is presented to the reader in some form (an equation, a table, figures, written prose, etc.) followed by a series of Critical Thinking Questions which lead the student to the development of a particular concept or idea. Whenever possible, data are presented *before* a theoretical explanation, and the Critical Thinking Questions lead the student through the thought processes which results in the construction of a particular theoretical model. This is what makes this book a *guided inquiry*. We have tried to mimic the scientific process as much as possible throughout this book. Students are often asked to make predictions based on the model that has been developed up to that point, and then further data or information is provided which can be compared to the prediction. It is important that these predictions be made BEFORE proceeding to get the full appreciation and benefit of this way of thinking. In this way, models can be confirmed, refined, or refuted, using the paradigm of the scientific method. The philosophical and pedagogic basis for this approach is described in a recent article.[2]

In addition to the student edition of *Physical Chemistry: A Guided Inquiry*: *Thermodynamics*, we have written several other documents to assist you in your endeavor to make the students better learners:

- The Instructor's Edition of *Physical Chemistry: A Guided Inquiry*: *Thermodynamics*. Available online rather than in printed form, this book is identical to the student edition except that it has answers to all of the focus questions and critical thinking questions (CTQs) in shaded boxes directly under the CTQs. Our answers are not the only answers or even the best answers. Rather, they serve to alert you as to the type of response we are expecting.
- The Solutions Manual for *Physical Chemistry: A Guided Inquiry*: *Thermodynamics*. This book contains worked-out solutions for all of the exercises and problems at the end of each ChemActivity. The solutions manual may be purchased by students.
- The Instructor's Guide for *Physical Chemistry Thermodynamics*. This online guide gives suggestions for how to organize the course, how to assign groups and roles, how to interact with groups, and how to respond to student questions. It also contains a table that correlates each ChemActivity to chapters in several other physical chemistry textbooks, the major concepts covered in each ChemActivity, the approximate time for each ChemActivity, and sample quizzes.

---

[1] Johnson, D. W. ; Johnson, R. T. *Cooperative Learning and Achievement.* In Sharon, S. (Ed.), *Cooperative Learning: Theory and Research*, pp 23-37, New York: Praeger.
[2] Spencer, J. N. *J. Chem. Educ.* **1999**, *76*, 566.

For information on accessing the online supplements (the Instructor's Edition and Instructor's Guide) please contact your Houghton Mifflin sales representative. You can find out who your representative is by visiting Houghton Mifflin's web site at college.hmco.com.

James N. Spencer   james.spencer@fandm.edu   Franklin & Marshall College
Richard S. Moog   richard.moog@fandm.edu   Chemistry Department
John J. Farrell   Lancaster, PA 17604

To the Student

Physical chemistry is a subdiscipline of chemistry that encompasses a quantitative study of the physical properties of chemicals and chemical reactions. For about a century the standard topics of physical chemistry have been thermodynamics (heat flow) and kinetics (rates of chemical reactions). These topics are generally concerned with the macroscopic properties of chemicals and chemical reactions. During the past 40 years or so, physical chemists have placed an increasing emphasis on understanding chemicals and chemical reactions at the molecular level—analysis of virtually one atom or molecule rather than a large collection of molecules. On the theoretical level this analysis is called quantum mechanics. On the experimental level this analysis is typically achieved by some form of spectroscopy.

This book about physical chemistry is *not* a textbook. This book is *not* a study guide. This book is a *guided inquiry*. Specifically, this book is a collection of group activities (each group has three or four students) that are to be accomplished in the presence of a mentor (instructor). Each group activity has one or more *models* (data, prose, or figures that represent the core of some chemical concept) followed by a series of *critical thinking questions* (CTQs). Systematically working through the CTQs in groups is essential for three basic reasons:

- Explaining concepts to other members of your group not only helps *their* understanding, it broadens *your* understanding. Instructors often have an exceptional understanding of the material they teach. One of the reasons for this depth of understanding is that teachers are constantly explaining concepts and exchanging ideas. Research has shown that this sort of verbal communication is a very important aspect of the learning process. Furthermore, it is often the case that someone who has just learned a concept is a better communicator of the concept to a novice than someone who is very familiar with the concept.

- Learning to ask questions that clearly and concisely describe what you do not understand is an important skill (not only in this and other courses but in all aspects of your life). It is a skill that improves with practice. When you do not receive the answer you expected from the other members of your group you may realize that the failing was in your question. You will learn how to ask better questions from your mistakes, from your mentor, from other members of your groups, and from reading the CTQs in the book.

- Groups (teams) have become essential to identifying, defining, and solving problems in our society. It is important that we learn how to be active and productive members of a group. If a member of your group is not contributing, it is your responsibility to help that member to become more productive. If a member of your group is over-contributing (thereby slowing the progress of the group) it is your responsibility to help that member to become more productive. Furthermore, as a member of the group you will have a role to play (manager, recorder, technician, and so on). Each role has a function important to the success of the group. Understanding the roles and dynamics of a group and how to change the dynamics of a group is a skill that can be transferred to many real-life situations.

We have found the use of these methods to be a more effective learning strategy than the traditional lecture, and the vast majority of our students have agreed. We hope that you will take ownership of your learning and that you will develop skills for lifelong learning. No one else can do it for you. We wish you well in this undertaking.

If you have any suggestions on how to improve this book, please write to us.

James N. Spencer   james.spencer@fandm.edu   Franklin & Marshall College
Richard S. Moog   richard.moog@fandm.edu   Chemistry Department
John J. Farrell   Lancaster, PA 17604

Acknowledgments

This book is the result of the innumerable interactions that we have had with a large number of stimulating and thoughtful people.

- Thanks to the following faculty members who have used some of our preliminary materials and have provided helpful feedback: Renee Cole (Central State University); Jeffrey Kovac (University of Tennessee); Ken Morton (Carson-Newman College); Susan Phillips (University of Pennsylvania); Marty Perry (Ouachita Baptist University); Clayton Spencer (Illinois College); George Shalhoub (La Salle University).

- Thanks to Richard Stratton, Houghton Mifflin Company, who had the vision and courage to bring a new educational pedagogy to college-level students. Thanks to Danielle Richardson, Katherine Greig, and Alexandra Shaw, all of Houghton Mifflin Company, for their help and encouragement throughout the entire process.

- Special thanks to Dan Apple, Pacific Crest Software, for taking us to this untravelled path and for pushing during the first two years of the journey. The Pacific Crest Teaching Institute we attended provided us with the insights and inspiration to convert our classroom into a fully student-centered environment.

- We also appreciate the support and encouragement of the many members of the Middle Atlantic Discovery Chemistry Project, who have provided us with an opportunity to discuss our ideas with interested, stimulating, and dedicated colleagues.

- Thanks to Carol Strausser, Franklin & Marshall College, for typing the photo-ready copy and for having sufficient patience to work with us through the editing and reediting.

- A great debt of thanks is due our students in Physical Chemistry at Franklin & Marshall College these past five years. Their enthusiasm for this approach, patience with our errors, and helpful and insightful comments have inspired us to continue to develop as instructors, and have helped us to improve these materials immeasurably.

Contents

Gases (I)

Focus Question: **What is a typical speed for a N_2 molecule in this room?**

a) 50 m/s   b) 500 m/s   c) 1500 m/s

Information

Long before the development of quantum mechanics, the kinetic molecular theory of gases had addressed such questions as:

- How fast do molecules of a gas move?
- What is the size of a molecule?
- How strongly are the molecules of a gas attracted to one another?

It had been established that the kinetic energy, *KE*, of a mole of gas is:

$$KE = 3/2 \, RT \, .$$

Model 1: The Kinetic Molecular Theory (KMT) of Gases.

- Any ordinary-sized, or macroscopic, sample of a gas contains a large number of particles. These particles can be treated as point particles (they have no volume).

- The particles move randomly in straight-line, continuous motion. Not all particles have the same speed; some particles are moving slowly, others are moving quickly.

- There are no attractive or repulsive forces between the particles.

- The molecules collide with each other and the walls of the container. These collisions are perfectly elastic (no translational energy is lost as a result of a collision).

- Pressure is caused by collisions of the particles with the walls of the container.

Critical Thinking Questions

1. According to Model 1, why does the pressure of a gas increase when the number of particles increases at a fixed volume?

2.   A gas that has all of the properties of the kinetic molecular theory is called an **ideal gas**. Predict which gas, helium or carbon tetrachloride, exhibits behavior closer to that of an ideal gas. Explain.

3.   Why does a collision with the wall of the container cause pressure?

Information

As stated in Model 1, the collisions of the particles with the walls of the container causes pressure. More explicitly, when a single molecule collides with a wall there is a change in momentum of the molecule. As shown in Figure 1, when a molecule collides with the wall, the momentum change, Δp, is given by $\Delta p = \Delta(mv)$,

$$\Delta p = (mv)_{after} - (mv)_{before} = mv_y - (-mv_y) = 2\,mv_y$$

when v_y is the component of velocity in a chosen direction.

A force is a change of momentum with respect to time:

$$f = \Delta p / \Delta t = 2\,mv_y / \Delta t$$

Figure 1.   One molecule causes pressure by collision with a wall of the container.

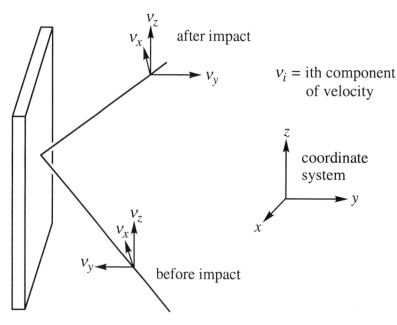

Pressure is force/area. Thus, the pressure caused by this one particle depends on the following factors: the mass of the particle; the velocity of the particle (which determines the momentum of the particle and how often it collides with the walls); the dimensions of the container (the larger the container, the fewer collisions with the walls).

In a collection of particles, all particles do not move with the same speed. The average speed, $\bar{v}$, is given by:

$$\bar{v} = \frac{v_1 + v_2 + \ldots + v_N}{N} = \frac{1}{N} \sum_{i=1}^{N} v_i$$

Model 2: The Kinetic Energy of a Particle.

$$\text{kinetic energy of one particle} = ke_i = \frac{1}{2} m v_i^2 \tag{1}$$

where k is Boltzmann's Constant.

Because the kinetic energy of a particle depends on the square of the velocity, a useful quantity is the average of the square of the velocity:

$$\overline{v^2} = \frac{1}{N} \sum_{i=1}^{N} v_i^2$$

It can be shown[1] that the pressure caused by N particles with mass m, each particle with a velocity v_i, in a cubic container of edge length l and volume of l^3, is

$$P = \frac{1}{3} N \frac{m\overline{v^2}}{V} \quad \text{or} \quad PV = \frac{1}{3} Nm\overline{v^2}$$

Because the number of moles, n, is equal to N/N_A, where N_A is Avogadro's number,

$$PV = \frac{1}{3} n N_A m \overline{v^2} \tag{2}$$

---

[1]This derivation is given in most standard Physical Chemistry texts in the chapter on gases.

Critical Thinking Questions

4. Show that:

   the kinetic energy of N particles $= \frac{1}{2} N m \overline{v^2}$

5. Show that:

   the kinetic energy of one mole of particles $= KE = \frac{1}{2} N_A m \overline{v^2}$ (3)

6. Use equations (2) and (3) to derive equation (4).

$$PV = \frac{2}{3} n\, KE$$ (4)

Information

The ideal gas law was formulated from experimental data. Many gases closely follow the ideal gas law over a wide (but limited) range of temperatures and pressures.

$$PV = nRT$$ (5)

Critical Thinking Questions

7. According to equation (5), for one mole of gas does the quantity *PV* depend on the type of gaseous molecule (for example: H_2, Ar, or methane) at a given temperature? Explain.

8. Keep in mind your answer to CTQ 7. According to equation (2), does the quantity $\frac{1}{3} n\, N_A\, m\overline{v^2}$ depend on the type of gaseous molecule (for example: H_2, Ar, or methane) at a given temperature? Explain.

9. Keep in mind your answer to CTQs 7 and 8. According to equation (3), for one mole of gas does the quantity $\frac{1}{2}\, N_A\, m\overline{v^2}$ depend on the type of gaseous molecule (for example: H_2, Ar, or methane) at a given temperature? Explain.

10. Assuming ideal behavior, the kinetic energy of one mole of H_2 is the same as the kinetic energy of one mole of methane (both at the same temperature). Explain why this is consistent with your answers to CTQs 7, 8, 9? However, the mass of an H_2 molecule is much less than the mass of a methane molecule. How can these facts be reconciled?

Exercise

1. Equal numbers of moles of H_2 and He gases at the same temperature are kept in the same size containers. Assume that H_2 and He obey the ideal gas law. a) In which container is the pressure greater? b) In which container is the average kinetic energy of the particles lower? c) In which container are the particles moving faster? d) In which container is the frequency of collisions with the walls the greatest?

Model 4: The Speeds of Particles.

A typical speed for particles is the root-mean-square (rms) speed.

$$\sqrt{\overline{v^2}} = \sqrt{\frac{3RT}{M}} = \text{root-mean-square speed (m s}^{-1}) \tag{6}$$

M is the molar mass (kg mol^{-1})
R is the gas constant (8.314 J K^{-1} mole^{-1})
T is the temperature (K)

Critical Thinking Questions

11. Derive $KE = \frac{3}{2}RT$ using equations (4) and (5).

12. Derive equation (6) above using equations (3), (4), and (5).

13. According to equation (6), what effect does an increase in temperature have on the rms speed of the gaseous particles?

14. Which gas has the greatest rms molecular speed at 298 K: H_2 or CH_4?

Exercises

2. Show that when the units given for the right-hand-side of equation (6) are used that the calculated velocity will have units m s^{-1}.

3. Calculate the root-mean-square speed for H_2 at 298 K and for methane at the same temperature.

4. a) Which has the greater root-mean-square speed—$^{235}UF_6$ or $^{238}UF_6$? b) If the container that holds a mixture of both gases, $^{235}UF_6$ and $^{238}UF_6$, has small holes, which gas would escape more quickly? (Note. The natural abundance of ^{235}U is 0.7%; to make the first atom bomb scientists need an abundance of about 2%. The small differential in diffusion rates of $^{235}UF_6$ and $^{238}UF_6$ was used to increase the abundance of ^{235}U to a fissionable level.)

5. A very small gas bubble seen in a microscope has a diameter of 1 μm. If the pressure is one bar and the temperature is 25 °C, calculate the number of gas molecules in the bubble.

6. (a) Calculate the pressure (in Pa and bar) of 10^{23} gas particles, each with a mass of 10^{-25} kg, in a 1-L container if the rms speed is 100 m/s. (b) What is the total kinetic energy of these particles? (c) What is the temperature inside the container?

7. (a) A 2.00-L gas bulb contains 4.07×10^{23} N_2 molecules. If the pressure is 3.05 bar, what is the rms speed of the N_2 molecules? (b) What is the temperature?

8. Calculate and compare the the average gas-phase translational energies and rms speeds of He(g) and H_2(g) at 20°C.

9. Assume that He is an ideal gas. (a) Calculate the rms speeds of He at 10, 100, and 1000 K in meters per second and in miles per hour. (b) What values would be obtained if the pressure was 10^{-10} bar?

10. The following values are found for the speed of sound in air:

°C	20	100	500	1000
Speed (m/s)	344	386	553	700

Compare these values to the rms speeds of dinitrogen molecules at these temperatures.

Model 5: The Maxwell Distribution of Molecular Speeds.

Not all gas particles within a collection of particles (at constant temperature) have the same velocity—some particles are moving slowly, others are moving rapidly. The problem of how to determine the most probable distribution of speeds was solved by J. Clerk Maxwell. Starting with the following assumptions,

- The probability of a molecular state depends only on the energy of the molecular state.

- The same probability distribution applies for all kinds of molecules.

Maxwell's equation for the distribution of speeds is the following:

$$\text{fraction of molecules per unit speed interval} = 4\pi v^2 \left(\frac{m}{2\pi kT}\right)^{3/2} e^{-mv^2/2kT} \qquad (7)$$

where k (Boltzmann's constant) $= 1.38066 \times 10^{-23}$ JK^{-1}

Table 1. The Speeds of Gaseous N$_2$ Molecules.

speed of molecule (m/s)	fraction of molecules per unit speed interval at 300 K (s/m)	fraction of molecules per unit speed interval at 1000 K (s/m)
0	0	0
25	1.87×10^{-5}	3.08×10^{-6}
50	7.40×10^{-5}	1.23×10^{-5}
100	2.84×10^{-4}	4.85×10^{-5}
200	9.60×10^{-4}	1.85×10^{-4}
300	1.63×10^{-3}	3.82×10^{-4}
400	1.96×10^{-3}	6.03×10^{-4}
500	1.84×10^{-3}	8.10×10^{-4}
600	1.43×10^{-3}	9.69×10^{-4}
700	9.39×10^{-4}	1.06×10^{-3}
800	5.28×10^{-4}	1.07×10^{-3}
1000	1.09×10^{-4}	9.15×10^{-4}
1100	4.07×10^{-5}	7.78×10^{-4}
1200	1.33×10^{-5}	6.28×10^{-4}
1300	3.84×10^{-6}	4.84×10^{-4}
1400	9.77×10^{-7}	3.56×10^{-4}
1500	2.20×10^{-7}	2.51×10^{-4}
1600	4.39×10^{-8}	1.69×10^{-4}
1700	7.77×10^{-9}	1.10×10^{-4}
1800	1.22×10^{-9}	6.81×10^{-5}
1900	1.70×10^{-10}	4.07×10^{-5}
2000	2.11×10^{-11}	2.34×10^{-5}

Ludwig Boltzmann developed a more general equation that dealt with energies.

Critical Thinking Questions

15. Consider a speed of 1800 m/s; at which temperature (300 or 1000 K) is the fraction of N_2 molecules per unit speed interval the greatest?

16. Consider a speed of 50 m/s; at which temperature (300 or 1000 K) is the fraction of N_2 molecules per unit speed interval the greatest?

17. There is a larger fraction of molecules moving at 1000 m/s when the gas temperature is 1000 K than when the gas temperature is 300 K. Does this make sense?

18. Why is there a smaller fraction of molecules moving at 50 m/s when the gas temperature is 1000 K than when the gas temperature is 300 K?

19. At 300 K, which speed has the greatest fraction of molecules per unit speed interval?

20. At 1000 K, which speed has the greatest fraction of molecules per unit speed interval?

21. Predict the speed that will have the greatest fraction of N_2 molecules per unit speed interval when the temperature is 600 K.

Exercises

11.   The most-probable speed, v_{mp}, is the speed for which equation (7) is a maximum. Show that $v_{mp} = \sqrt{\dfrac{2RT}{M}}$.

12.   What is the most-probable speed for N_2 at 300 K?

13.   It can be shown that the average speed, $\bar{v}$, is given by $\bar{v} = \sqrt{\dfrac{8RT}{\pi M}}$. What is the average speed for N_2 at 300 K?

14.   Use the data in Table 1 to prepare a plot of the fraction of molecules per unit speed interval vs. speed for N_2 gas at 300 K.   Indicate the most-probable speed, the average speed, and the root-mean-square speed on the graph. Plot the data at 1000 K on the same graph. Indicate the most probable speed, the average speed, and the root-mean-square speed on the graph.   Sketch the fraction of molecules per unit speed interval vs. speed for N_2 gas at 1200 K on the same graph.

15.   Which is greater—the average speed or the most-probable speed? Use the shape of the curve generated in Exercise 14 to explain why this is the case.

16.   Why is the root-mean-square speed greater than the average speed? [Hint. Take the average of the following three numbers: 1,3,5. Take the average of $1^2, 3^2, 5^2$.]

17.   The speed of space shuttle in orbit is about 18,000 mi per hr. Which is greater, the average speed of a N_2 molecule at 300 K or the speed of space shuttle in orbit?

G1A

The Maxwell-Boltzmann Speed Distribution Law

Focus Question: Are all the O_2 molecules in the room moving with the same kinetic energy? the same velocity?

Information

The Kinetic Molecular Theory assumes that a gas is comprised of a large number of particles that are small in comparison with the distances between them and the size of the container. These gaseous particles are in continuous random motion, and collisions between the molecules and the walls of the container are perfectly elastic, that is the kinetic energy before and after a collision is unchanged. It is assumed that each particle moves with a different kinetic energy. The mathematical relationship that shows how speeds are distributed among the molecules was derived by Maxwell and Boltzmann.

Model 1: The Distribution of Speeds for Nitrogen Gas at 300 and 1000 K.

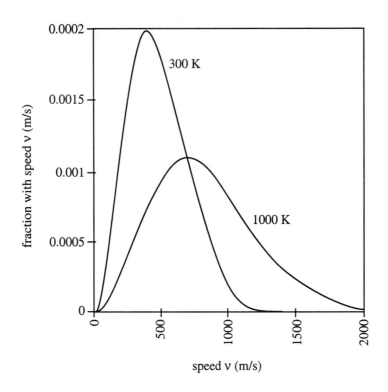

Critical Thinking Questions

1. Each of the velocity components v_x, v_y, and v_z can vary from $-\infty$ to $+\infty$. What is the range spanned by the speed v ?

2. The speed, v, of a particle is related to its velocity components by $v^2 = v_x{}^2 + v_y{}^2 + v_z{}^2$. If the speed of a particle is 500 m/s list three ways this speed could be distributed among the velocity components.

3. Plot the particle speed on the graph below as shown in the model for each set of velocity components selected in CTQ 2.

Information

The KMT model assumes that there is a function $f(v_x)$, that represents the fraction of all particles $dN(v_x)/N\, dv_x$ with velocity component v_x in the range from v_x to $v_x + dv_x$.

$$\frac{1}{N}\frac{dN(v_x)}{dv_x} = f(v_x)$$

or

$$\frac{dN(v_x)}{N} = f(v_x)\, dv_x$$

There are similar functions that describe the fraction of particles for velocity components v_y and v_z. Thus, because each function represents a probability and if the velocity components are independent, the probability is

$$F(v) = f(v_x)f(v_y)f(v_z)\,. \tag{1}$$

or

$$\frac{dN(v)}{N} = F(v)dv = f(v_x)f(v_y)f(v_z)\, dv_x dv_y dv_z\,. \tag{2}$$

Critical Thinking Questions

4. Take the natural logarithm of each side of eqn (1) and then differentiate the resulting equation with respect to v_x.

5. Use the chain rule to show that

$$\left.\frac{\partial \ln F(v)}{\partial v_x}\right)_{y,z} = \frac{d\ln F(v)}{dv} \left.\frac{\partial v}{\partial v_x}\right)_{y,z} = \frac{d\ln f(v_x)}{dv_x} \, . \tag{3}$$

6. If $v = \sqrt{v_x^2 + v_y^2 + v_z^2}$ find $\left.\dfrac{\partial v}{\partial v_x}\right)_{y,z}$.

7. Use the results of CTQ 6 and equation (3) to show that

$$\frac{1}{v}\frac{d\ln F(v)}{dv} = \frac{1}{v_x}\frac{d\ln f(v_x)}{dv_x} \, . \tag{4}$$

8. The left-hand side (LHS) of equation (4) is a function of v. What are the variables on the right-hand side of the equation?

9. If v_y is varied how does the LHS of equation (4) change? If v_z is varied how does the LHS of equation (4) change?

10. What can you conclude about the LHS of equation (4)? Is it a constant or does it vary with changing v_y, v_z or v_x?

11. If $\dfrac{1}{v}\dfrac{d\ln F(v)}{dv}$ = const = $-a$, find $F(v)$. Let A represent the constant of integration.

12. Show that

$$\frac{dN(v)}{Ndv_x dv_y dv_z} = Ae^{\frac{-av^2}{2}} = F(v) \ .$$

Model 2: The Speed is Independent of the Angle from the Origin $(v_x = v_y = v_z = 0)$.

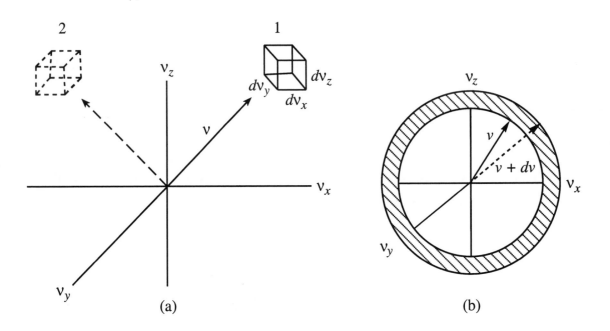

(a)   (b)

For a given speed, v, the probability is not dependent upon the direction from the origin. Thus the same number of particles can be found with velocities in the volume 1 above as in volume 2; i.e., the speed is independent of the angle from the origin. The annular volume (spherical shell) between v and $v + dv$ contains all the particles with components of velocity between $v_x + dv_x$, $v_y + dv_y$, and $v_z + dv_z$.

Critical Thinking Questions

13. The volume of a sphere is $Vol = 4/3 \, \pi r^3$ where r is the radius of the sphere. What is the volume of the annular region between the speed of v and $v + dv$?

14. Use the relationship in CTQ 12 and your answer to CTQ 13 to show that

$$\frac{dN(v)}{N} = Ae^{\frac{-av^2}{2}} \cdot 4\pi v^2 dv \ . \tag{5}$$

15. If the LHS of CTQ 14 is integrated over all space, the value must be equal to unity. Why?

Information

Eqn (5) contains two constants A and a. These constants can be determined by making use of previous relationships.

It can be shown that $A = \left(\dfrac{a}{2\pi}\right)^{3/2}$. [See Exercise 1.]

Critical Thinking Questions

16. In ChemActivity G1B, the average value theorem will be used to show that $a = m/kT$. Find A.

17. Use equation (5) to show that

$$\frac{dN(v)}{N} = 4\pi \left(\frac{m}{2\pi kT}\right)^{3/2} v^2 \, e^{\frac{-mv^2}{2kT}} \, dv.$$

18. Show that

$$F(v) = \left(\frac{m}{2\pi kT}\right)^{3/2} e^{\frac{-mv^2}{2kT}} .$$

$$= \left(\frac{m}{2\pi kT}\right)^{3/2} e^{\frac{-mv_x^2}{2kT}} \; e^{\frac{-mv_y^2}{2kT}} \; e^{\frac{-mv_z^2}{2kT}} .$$

19. From CTQ 18 find expressions for $f(v_x)$, $f(v_y)$, and $f(v_z)$.

Exercises

1. Integrate equation (5) to show that

$$A = \left(\frac{a}{2\pi}\right)^{3/2} .$$

Recall that $\displaystyle\int_0^\infty x^2 e^{-b^2x^2}\, dx = \frac{\sqrt{\pi}}{4b^3} .$

2. How is the Boltzmann constant k related to the gas law constant R? Show that $e^{\frac{-mv^2}{2kT}} = e^{\frac{-Mv^2}{2RT}}$ where m is the mass of a particle and M is the mass of Avogadro's number of particles.

3. Roughly plot the one-dimensional velocity distribution against v_x at two different temperatures.

4. Plot the three-dimensional velocity distribution at two different temperatures.

5. What is the effect of changing the mass at a given temperature on the plot of the three-dimensional function?

6. Show that the velocity distribution in two dimensions may be written

$$\frac{dN(v)}{N} = \left(\frac{m}{2\pi kT}\right) \bullet 2\pi \bullet e^{\frac{-mv^2}{2kT}} \, v \, dv.$$

Hint: In two dimensions what is the annular area?

7. If $\varepsilon = 1/2 \, mv^2$ show that in three dimensions

$$\frac{dN(\varepsilon)}{N} = 2\pi \left(\frac{1}{\pi kT}\right)^{3/2} \varepsilon^{1/2} \, e^{-\varepsilon/\kappa T} \, d\varepsilon \ .$$

ChemActivity G1B

Average Values

Focus Question: Two molecules move at a speed of 1000 ms^{-1} and one at 1200 ms^{-1}. What is the average speed of the molecules?

Model 1: A Basketball Team with Five Players.

Height of Player	Position
6.5 ft	Guard
6.8 ft	Forward
6.5 ft	Guard
6.8 ft	Forward
7.0 ft	Center

Critical Thinking Questions

1. What is the average height, $\overline{h}$, of the team?

2. How, in general, could the average value of any parameter, r, for a group of objects, having N_i individual values of the parameter be determined? Give a written as well as a mathematical expression.

Model 2: Another Method for Calculation of an Average.

$$\overline{r} = \sum_i \frac{r_i N_i}{N} = \frac{1}{N} \int_{\substack{all \\ values}} r\,dN$$

Critical Thinking Questions

3. Define all terms in the model.

4. According to the Maxwell-Boltzmann speed distribution law, how many particles dN in a gas have a particular speed? Give only the mathematical representation.

5. Set up the mathematical relationship that will allow the calculation of the average speed, $\overline{v}$, from the model.

6. Standard tables of integrals give the following:

$$\int_0^\infty x^3 e^{-\beta x^2}\,dx = \frac{1}{2\beta^2}$$

Show that
$$\overline{v} = \left(\frac{8kT}{\pi m}\right)^{1/2}$$

Exercises

1. Find an expression for the root-mean-square speed $\sqrt{\overline{v^2}}$. You will need the standard integral:

$$\int_0^\infty x^4 e^{-\beta x^2} \, dx = \frac{\sqrt{\pi}}{2} \cdot \frac{3}{4} \frac{1}{\beta^{5/2}}$$

2. Show that the average speed of particles moving the positive x-direction only, v_x, is given by $|\overline{v_x}| = \dfrac{\overline{v}}{2}$.

$$\int_0^\infty x e^{-\beta x^2} \, dx = \frac{1}{2\beta}$$

Note that the velocity distribution function is needed in one dimension only.

3. For $O_2(g)$ at 298 K compare $\sqrt{\overline{v^2}}$, $\overline{v}$, and v_{mp}.

4. A single molecule moving with velocity component, v_x, encounters walls, normal to x, separated by distance, a, with frequency, $v_x/2a$. The collision frequency per unit area is $v_x/2V$. For N independent molecules show that the collision frequency, ν, may be given by

$$\nu = \frac{N|\overline{v_x}|}{2V}$$

where $|v_x|$ is the absolute value of the component of velocity in the x direction.

Then show that

$$\nu = \frac{N}{V} \frac{\overline{v}}{4}$$

and that

$$\nu = \frac{P}{\sqrt{2\pi kTm}}. \qquad P = \text{pressure}$$

5. The vapor pressure of water at 30°C is 31.8 Torr. How many water molecules per second strike a surface area of 10 cm^2 at 30°C in a beaker of water?

6. Two identical flasks are filled with Ne and Ar gas, respectively. The Ne flask is at twice the temperature but half the pressure of the Ar flask. What is the ratio of wall collision frequencies?

ChemActivity G2

Gases (II)

Focus Question: How does the volume of one mole of $CH_4(g)$ at 300 bar and 298 K compare to the volume of one mole of ideal gas at 300 bar and 298 K?

    a) volume of CH_4 is greater

    b) volume of CH_4 is less

    c) volume of CH_4 and ideal gas are the same

As mentioned previously, the kinetic molecular theory enabled scientists to address, for the first time, three important questions concerning the properties of gaseous molecules:

- How fast do gaseous molecules move?
- What is the size of a gaseous molecule?
- How strongly are gaseous molecules attracted to each other?

Our ability to estimate the size of molecules and the strength of the attractive forces between molecules was made possible by the insight of the Dutch chemist J. D. van der Waals and by our ability to measure deviations (in pressure and volume) from the ideal gas law.

Model 1: The Ideal Gas.

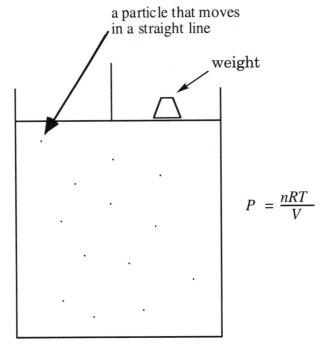

P is the gas pressure (bar)
n is the number of moles of gas
R is the gas constant, 0.083144 L bar K^{-1} mol^{-1}
T is the gas temperature (K)
V is the volume of the container (L)

Critical Thinking Questions

1. According to the ideal gas equation, if the number of moles of gas and the temperature are held constant, what will happen to the pressure of the gas when the volume is decreased?

2. What is the value of PV tor one mole of an ideal gas at 25.00°C and one bar?

3. What is the value of PV for one mole of an ideal gas at 25.00°C and 0.1000 bar?

4. What is the value of *PV* for one mole of an ideal gas at 25.00°C and 100.0 bar?

5. Make a sketch of *PV* vs. *P* for one mole of an ideal gas at 25.00°C.

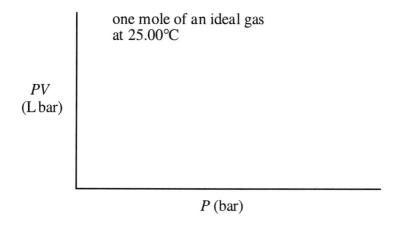

6. If *P* = 1.000 bar and *T* = 25.00°C, what is the volume of one mole of an ideal gas?

Information

The compressibility factor, *Z*, of a gas is defined as follows:

$$Z = \frac{P\overline{V}}{RT}$$ where $\overline{V}$ is the volume of one mole of gas (or molar volume).

Critical Thinking Questions

7. What is the value of *Z* for an ideal gas? What are the units of *Z*?

8.   Make a sketch of Z vs. P for one mole of an ideal gas at 25.00°C.

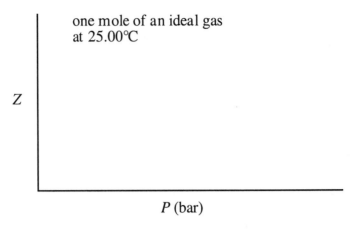

Information

The compressibility factor, Z, can be thought of as a measure of the deviation of a real gas from ideality.

Figure 1. Variation of compressibility factor of various gases with pressure at 298 K.[1]

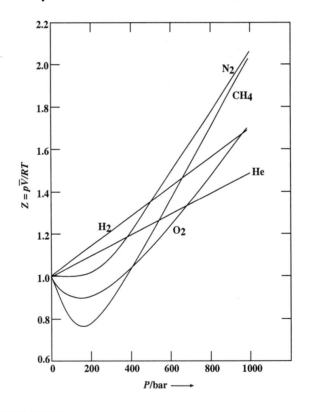

---

[1] R.A. Alberty, R.J. Silbey, *Physical Chemistry*, John Wiley & Sons, Inc., NY, 1992. This material is used by permission of John Wiley & Sons, Inc.

Critical Thinking Questions

9. Add a sketch of Z vs. P for an ideal gas at 298 K to Figure 1. Explain why you drew the sketch as you did.

10. Based on the data presented in Figure 1, for real gases what is

    a) $\lim\limits_{P \to 0} Z$?

    b) $\lim\limits_{P \to 0} P\overline{V}$

11. Under what conditions of pressure do real gases behave most ideally?

12. If $Z > 1$, is $\overline{V}$ greater than, less than, or equal to $\overline{V}_{ideal}$? Explain your reasoning.

13. If $Z < 1$, is the gas "more compressible" or "less compressible" than an ideal gas? Explain your reasoning using grammatically correct English sentences.

14. One of the assumptions of the KMT of gases is that the particles have negligible volume. Is this assumption valid or invalid for a real gas?

Exercises

1. Assume that a certain gas is ideal. What volume will the gas occupy at a pressure of 0.0465 bar if its volume is 3.00 L at 11.0 bar and the temperature is held constant?

2. What volume will a gas which behaves ideally occupy at 0°C if its occupies a volume of 11.4 L at 100°C and the pressure remains constant?

3. Make a plot of P vs. V at 0°C for a sample of an ideal gas that has a volume of 1.00 L at 0°C and 1.00 bar. On the same graph, plot P vs. V for the same sample of gas at 100°C.

4. A sample of gas has a volume of 1.00 L at sea-level conditions of 1.00 bar and 20°C. Calculate the volume of this same sample of gas at 500 km above the Earth's surface, the thermosphere, where the pressure is approximately 1.5×10^{-11} bar and the temperature is about 1600 K.

5. Use the ideal gas law and the definition of density, $d = \dfrac{mass}{volume}$, to show that the density of an ideal gas is given by: $d = \dfrac{PM}{RT}$.

6. Assume that N_2 behaves ideally and calculate the density in grams per liter of N_2 at 20°C and 1.00 bar.

7. The density of a gas at 0.500 bar and 20°C is 0.6565 g/L. Assume that the gas is ideal and calculate the molecular mass of the gas.

8. Calculate the average molecular mass in the Earth's thermosphere (about 500 km) where the pressure is approximately 1.5×10^{-11} bar, the temperature is about 1600 K, and the density is about 2.2×10^{-12} g/L.

9. The Earth's atmosphere at about 500 km is composed primarily of He and about equal molar amounts of O and N_2. Qualitatively, are these data in agreement with your answer to Exercise 8? Comment on why the particular species given here are more reasonable than He, N and O_2.

10. A gas mixture contains 50.0 g of O_2 and 50.0 g of CO_2. Assume that these gas behave ideally and calculate the volume if the total pressure is 0.750 bar and the temperature is 20°C. What are the mole fractions and pressure fractions of each gas?

Model 2: Real Gases/Size Factor.

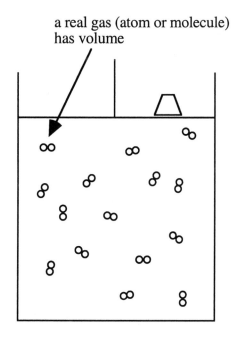

a real gas (atom or molecule)
has volume

Critical Thinking Questions

15. For an ideal gas, $P = \dfrac{nRT}{V}$. Assuming that n, R, V and T are held constant, will the pressure of the real gas depicted in Model 2 be smaller, the same, or larger than the value of P calculated from the ideal gas law equation ? Explain.

16. For an ideal gas, $P = \dfrac{nRT}{V}$. How would you modify this equation for a real gas to correct for the fact that molecules have volume?

Model 3: Real Gases/Size Factor.

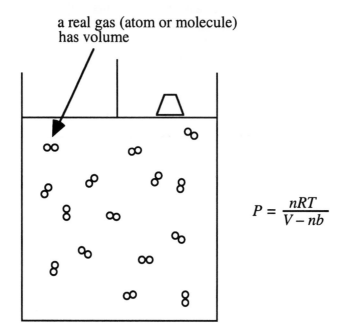

a real gas (atom or molecule)
has volume

$$P = \frac{nRT}{V - nb}$$

b is the excluded volume per mole of gas
$V - nb$ is the *effective* volume of the container

Critical Thinking Questions

17.   Assume that the volume of the container in Model 3 is 24.789 L and that the
      container holds one mole of a real gas. At 25°C, will the real gas depicted in Model
      3 have a greater pressure or a lower pressure than the ideal gas in Model 1?
      Explain.

18.   For methane, CH_4, $b = 0.0428$ L mol^{-1}. Use the equation in Model 3 to calculate
      the values of P, PV, and Z for one mole of methane at 25.00 °C and the volumes
      listed in the table below. (Record all values to four significant figures.)

Volume (L)	Pressure (bar)	PV (L bar)	Z
247.9	0.1000	24.79	1.000
123.99	0.2000	24.80	1.000
24.79	1.002		1.002
4.958		25.00	1.008
2.479	10.18	25.24	
0.2479			1.209

19.   In the table in CTQ 18, are the *PV* values lower or higher than those predicted by the ideal gas law? Explain.

20.   In the table in CTQ 18, are the *Z* values lower or higher than those predicted by the ideal gas law? Explain.

21.   Make a sketch of *Z* vs. *P* for the data in the table in CTQ 18.

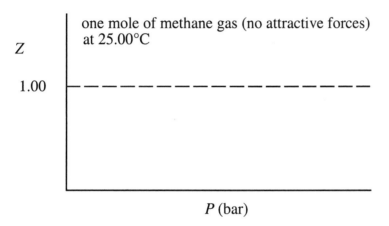

22.   One of the assumptions of the KMT of gases is that there are no attractive or repulsive forces between the molecules. Is this assumption valid or invalid for a real gas? Explain.

23.   What effect will attractive forces have on the pressure of a gas?

24.   For an ideal gas, $P = \dfrac{nRT}{V}$, and for a gas whose molecules have volume, $P = \dfrac{nRT}{V - nb}$. How would you modify this latter equation for a real gas to correct for the fact that there are attractive forces between molecules?

Model 4: Real Gases/Attractive Forces and Size Factors.

the particles of a real gas have volume
and are attracted to one another

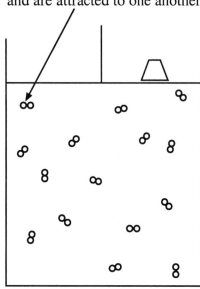

van der Waals' equation

$$P = \frac{nRT}{V - nb} - \frac{an^2}{V^2}$$

b is the excluded volume per mole of gas
$V-nb$ is the *effective* volume of the container
a is a measure of the strength of the attractive forces between particles
  a = 2.29 bar L^2 mol^{-2} for CH$_4$(g)
  a = 5.52 bar L^2 mol^{-2} for H$_2$O(g)

Critical Thinking Questions

25. Why is van der Waals' a value for H$_2$O(g) larger than the Van der Waals' a value
for CH$_4$(g)?

26. Assume that the volume of the container in Model 4 is 24.789 L and that the
container holds one mole of a real gas. At 25°C, will the real gas depicted in Model
4 have a greater pressure or a lower pressure than the gas in Model 3? Explain.

Data

Table 1. *PV* values calculated from the van der Waals equation for one mole of methane at 25°C.

Volume (L)	Pressure (bar)	PV (L bar)	Z
247.9	0.09998	24.78	1.000
123.99	0.1999	24.78	1.000
24.79	0.9980	24.74	0.998
4.958	4.950	24.54	0.990
2.479	9.803	24.30	0.980
0.2479	83.60	20.72	0.836

27.   In Table 1, are the Z values lower or higher than those predicted by the ideal gas law? Explain.

28.   Make a sketch of Z vs. P for the data in Table 1.

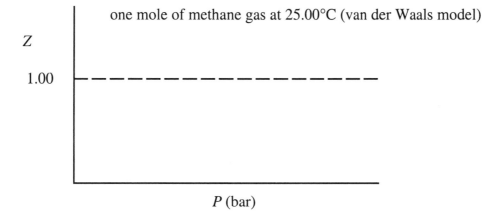

29.   Which factor appears to be the most important factor (at these temperatures and pressures) in causing deviations from ideality: the volume of the molecules or the attractive forces between the molecules? Explain.

Information

The van der Waals equation gives a better description of a real gas than the ideal gas equation. The values of a and b can be determined by several methods. One method is to plot Z vs. P and to adjust a and b to obtain the best fit.

A number of other two-parameter equations have been developed. The Redlich-Kwong equation, given below, is regarded as the best of these equations, but it is more complex than the van der Waals equation.

$$\left[P + \frac{n^2 a}{T^{1/2} V(V+nb)}\right](V-nb) = nRT$$

Exercises

11.   Carefully examine the following graph:

Z vs. P for a Real Gas at Several Temperatures

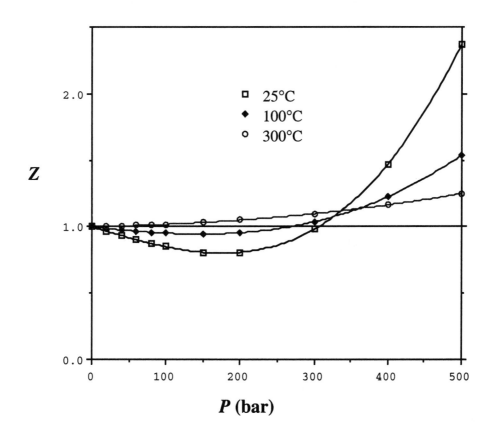

P (bar)

a.   At what pressures does this real gas tend to exhibit ideal gas behavior: low pressure or high pressure? Explain

b.   At what temperatures does this real gas tend to exhibit ideal gas behavior: low temperature or high temperature? Explain.

c.   Which of the following statements is most correct:
   • Gases tend to be ideal at high pressures and high temperatures.
   • Gases tend to be ideal at low pressures and low temperatures.
   • Gases tend to be ideal at low pressures and high temperatures.
   • Gases tend to be ideal at high pressures and low temperatures.

12.   In general, Z tends to be larger than 1 at very high pressures (greater than 350 bar for methane). Which van der Waals' factor seems to be responsible for this behavior— *a* or *b*? Explain.

13.   In general, Z tends to be smaller than 1 at moderately high pressures (around 150 bar for methane). Which van der Waals' factor seems to be responsible for this behavior— *a* or *b*? Explain.

14. A pressure of 43.6 bar is required reduce the volume of one mole of ammonia to 0.935 L at 200°C. (a) What pressure would be required according to the ideal gas law? (b) What pressure would be required according to van der Waals' equation? [For ammonia, $a = 4.25$ bar L^2 mol^{-2} and $b = 0.0374$ L mol^{-1}.]

15. For each pair of particles, which is expected to have the larger van der Waals' a value?
    a.   benzene or chlorobenzene
    b. H_2 or Cl_2
    c. CH_3OCH_3 or CH_3CH_2OH
    d.   He or Ar
    e. CH_4 or NH_3
    f. $N(CH_3)_3$ or $HN(CH_3)_2$

16. For each pair of particles, which is expected to have the larger van der Waals' b value?
    a.   benzene or chlorobenzene
    b. H_2 or Cl_2
    c. CH_3OCH_3 or CH_3OH
    d.   He or Ar
    e. CH_4 or CH_3NH_2
    f. $N(CH_3)_3$ or $HN(CH_3)_2$

17. Two gases have the following *P-V* relationships at 300 K:

	$P = 1.00$ bar	$P = 10.0$ bar
one mole of Gas A	$V = 24.80$ L/mole	$V = 2.35$ L/mol
one mole of Gas B	$V = 24.95$ L/mole	$V = 2.51$ L/mol

    One of these gases is Ne and the other is NH_3. Identify each and explain your reasoning.

18. When 0.1 mol of HCl gas is placed in a 10 liter box at 25°C, the pressure is about the same as that observed when 0.1 mol of NH_3 is placed in an identical box at the same temperature. However, when 0.05 mol of HCl and 0.05 mol of NH_3 are mixed in a 10 liter box at 25°C, a noticeably lower pressure is observed. Explain this behavior.

19. One of the lasers commonly used to produce blue and green light in laser light shows is the Ar^+ ion laser, which operates using gaseous Ar^+ ions. Consider a gas of these Ar^+ ions. Would you expect the compressibility factor for this system to be greater than, less than, or equal to 1? Explain your reasoning.

Problem

1. We have seen that gases tend to be ideal at high temperature and low pressure. One can also state that for a real gas:

    as $P \rightarrow 0$, $PV \rightarrow nRT$

    (a) For a real gas, as $P \rightarrow 0$, $d/P \rightarrow$? [d is density] (b) Use the data below for CO_2 at 0°C to prepare a graph of d/P vs. P. (c) Extrapolate the data to $P = 0$ bar. If the molecular mass of CO_2 is given as 0.044009 kg/mol, determine the value of the universal gas constant, R.

P(bar)	1.0000	0.75000	0.50000	0.35000	0.20000
d (g/L)	1.9509	1.4607	0.97217	0.67982	0.38807

ChemActivity T1

Work

Focus Question: For the following process is there any work done? If not, why not? If so, is work done by or on the system?

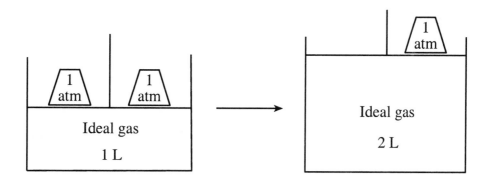

Information

Work is defined by equation (1):

$$\text{work} = \text{force} \times \text{distance} \tag{1}$$

For many chemical applications it is more convenient to treat work as pressure operating through a volume, which is an equivalent construction.

Model 1:   A Piston-Cylinder Containing a Gaseous System.

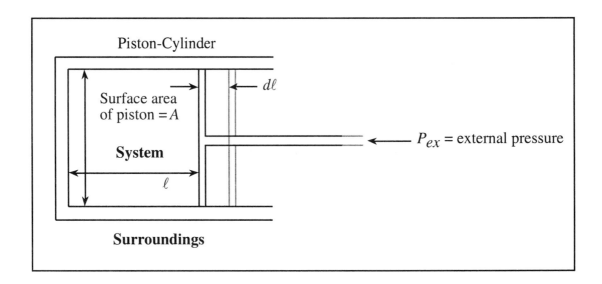

$$\frac{\text{force}}{\text{area}} = \text{external pressure} \quad ; \quad \frac{f}{A} = P_{ex} \tag{2}$$

When the system is compressed by an infinitesimal amount, $d\ell$, by a constant force, f, then an infinitesimal amount of work, dw, is done on the system.

By convention, work done *on the system* is considered to be positive; when the system does work *on the surroundings*, work is negative.

Critical Thinking Questions

1. If the gas in the system expands, is the work positive or negative? Explain.

2.   If the gaseous system is compressed, is the work positive or negative? Explain.

3.   Provide an expression relating the infinitesimal work, dw, to the force, f, and the infinitesimal distance, $d\ell$. Make sure that the sign convention described above is followed.

4.   Use equation (2) to modify the expression from CTQ 3 above to express the infinitesimal work, dw, in terms of external pressure, P_{ex}, and the infinitesimal change in volume, dV. Is the sign convention described above followed in your expression?

Model 2: Work Associated With Moving a Piston.

When the piston in Model 1 moves from an initial position ℓ_i to a final position ℓ_f, there is a change in the volume of the system. The initial volume is V_i and the final volume is V_f.

The total amount of work can be obtained by integrating both sides of the expression obtained in CTQ 4.   Work involved in the process is **defined** below:

$$\text{work} \ = \ w \ = \ \int_{\ell_i}^{\ell_f} dw \tag{3}$$

Critical Thinking Questions

5.   In order to integrate the equation from CTQ 4 it is necessary to determine the limits of integration.

   a)   When the piston is at ℓ_i, what is the volume?

b)   When the piston is at ℓ_f, what is the volume?

c)   Derive an equation describing the relationship between work, w, and the initial and final volumes of a gaseous sample for a process which takes place under constant P_{ex}. Hint: Use Model 2 and equation (3) and your answer to CTQ 4.

Information

There are a number of different circumstances under which gaseous systems can undergo changes. Examples of several of these are:

- Expansion against zero pressure
- Processes occurring at constant volume
- Processes occurring at constant (nonzero) pressure
- Processes for which all intermediate states of the process are in equilibrium. This is known as a reversible process.

Chemists find it useful to be able to calculate the work associated with these types of changes in gaseous systems.

Critical Thinking Questions

6.   Calculate the work done in an expansion against zero pressure.

7.   Calculate the work done in a constant volume process.

8.   Provide an expression for the amount of work done in a constant external pressure process in which there is a change of volume ΔV.

Model 3:   Work Associated With Moving a Piston when P_{ex} is Equal to P_{int}—A Reversible Process.

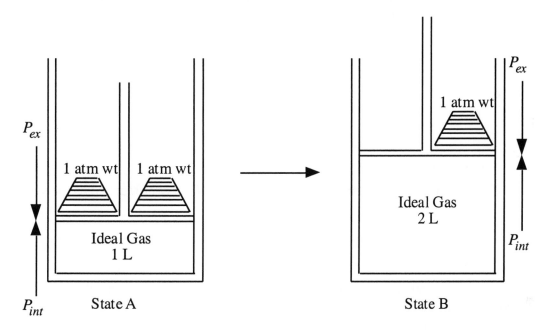

State A   State B

For each of states A and B, the external and internal pressures are equal; that is, the systems are at equilibrium. Both states contain the same number of moles of gas. One possible way to get from state A to state B would be to remove one weight.

Critical Thinking Questions

9.   Based on the descriptions of states A and B in Model 3, are the two systems at the same temperature? If not, which one is at a higher temperature? Explain your reasoning.

10. What is the relationship, at all times, between P_{ex} and P_{int} for a reversible process?

11. Consider the removal of one of the 1 atm weights from the piston in state A:

   a)   Is the system still at equilibrium at the moment that the weight is removed? When the weight is removed, what will happen?

   b)   Does the process you described in part a) occur against a constant external pressure? If so, what is this pressure?

   c)   When state B is reached, how has the pressure of the gas in the cylinder changed as a result of the process?

   d)   Is an equilibrium state (one in which $P_{ex} = P_{int}$) ever reached as the system changes from state A to state B?

12. Suggest a hypothetical way to get from state A to state B via a reversible process - that is, such that $P_{ex} = P_{int}$ at all times? (Hint: You need not use the specific weights shown in Model 3.)

13. Use the ideal gas equation and your answer to CTQ 4 to obtain an expression relating dw to dV for a reversible process for an ideal gas. This expression should not contain the variable P.

14. Integrate both sides of your answer to CTQ 13 to obtain an expression relating work (w) to the initial and final volumes for an isothermal (constant T) reversible process.

15. Why would CTQ 14 be much more difficult if the process were NOT isothermal?

Exercises

1. Consider the process below:

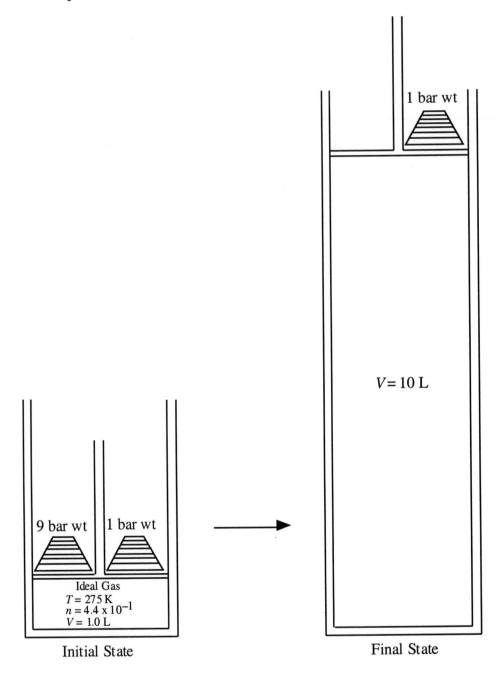

9 bar wt   1 bar wt

Ideal Gas
$T = 275\,K$
$n = 4.4 \times 10^{-1}$
$V = 1.0\,L$

Initial State

1 bar wt

$V = 10\,L$

Final State

a)   Calculate the work if the process is carried out reversibly and isothermally.

b)   Calculate the work if the process is carried out irreversibly in one step.

2. Consider the same process as in Exercise 1, but let the process go from right to left— that is, a compression.

    a)   Calculate the work if the process is carried out reversibly and isothermally.

    b)   Calculate the work if the process is carried out irreversibly in one step.

    c)   Compare the values obtained in a) and b) to those obtained in Exercise 1.

3. A gas obeys the equation of state:

    $$PV = RT + \alpha P \quad \text{where } \alpha \text{ is a constant}$$

    a)   Calculate the work if the gas is heated at constant pressure from T_1 to T_2.

    b)   Calculate the work if the gas is expanded isothermally and reversibly.

The First Law of Thermodynamics

Focus Question: A hot brick is placed into cold water in an isolated container. The final temperatures of the brick and the water are identical. What is the total energy change in this process:

a) positive

b) negative

c) zero

d) cannot determine without further information

Model 1: An Ideal Gas in a Piston-Cylinder with External Pressure from a Weight.

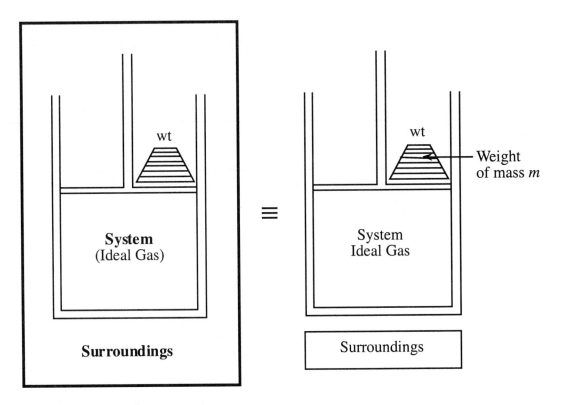

Surface area of piston = A
Force of weight = mg
Potential energy of weight = mgh
$\qquad\qquad$ g = acceleration of gravity
$\qquad\qquad$ h = height of the weight

In dealing with thermodynamic analysis it is often convenient to divide the universe into three components:

- the **system** - the chemical system (for example, a gas within a cylinder) of interest that is undergoing some kind of change

- the **mechanical surroundings** - that part of the universe that can undergo a mechical change, such as the raising or lowering of a weight, as a result of its interaction with the system.

- the **thermal surroundings** - everything else in the universe that might conceivably undergo a thermal change when the system changes.

The energy, U_{sys}, associated with the system is often of particular interest, as are changes in the energy of the system, ΔU_{sys}. In the above diagram, the external pressure source is a weight of mass m. The energy associated with this source is designated U_{wt}.

Information

The first law of thermodynamics states that energy is conserved: in any process, the total energy before and after the process has been carried out is the same. The first law also tells us that energy is additive. That is, the total energy is the sum of the energies of its parts. The first law of thermodynamics is a law of experience. It cannot be derived from general principles, but no experiment has been devised that contradicts this law. Thus, we accept this natural law as being of universal validity.

Critical Thinking Questions

1. What is a reasonable symbol to represent the energy of the surroundings?

2. Provide an expression relating the total energy of the universe, U_{tot}, to the energies of the system, the external pressure source, and the surroundings.

3. Provide an expression for the change in the total energy of the universe, ΔU_{tot}, in terms of the changes in energy of the system, surroundings, and external pressure source (U_{wt}).

4. Consider a situation in which the surroundings in Model 1 heat the system reversibly, causing it to expand. The position of the piston then increases by an amount Δh.

    a) Is work done in this process? If not, why not? If so, is this positive or negative work?

    b) If more heat is supplied by the surroundings, would you expect the magnitude of work to be increased, decreased, or unaffected? Why?

5. Recall that the potential energy of a mass in a gravitational field is given by $V = mgh$, where h is a measure of height.

   a)   If h increases what happens to V?

   b)   Assuming that the only energy changes for the mechanical surroundings are associated with changes in potential energy of the position of the weight, provide an equation for ΔU_{wt} in terms of m, g, and h.

   c)   Use your answer to part b) above, along with the relationship of pressure, force, and area to develop an expression for ΔU_{wt} in terms of the external pressure and the change in volume of the system.

6. Consider the process described in CTQ 4:

   a)   Is the external pressure constant for this process?

   b)   Is the pressure of the system, the internal pressure, constant?

   c)   Laboratory experiments are usually not carried out in a piston-cylinder with a weight attached, but in open vessels. What is the external pressure source in these experiments?

Model 2: Heat and Work.

There are two common ways the energy of a system can be altered:

- the system can gain or lose heat
- the system can do work or have work done on it

Both heat and work are manifest by changes in the surroundings and appear only during a change in state. **Heat**, q, is energy in transit, and as a result of heat exchange the surroundings will be hotter or colder. When the surroundings get colder, q is defined to be positive. Work, like heat, appears at the boundary of systems, and can be thought of as being equivalent to the raising or lowering of a weight in the surroundings.

Work is an exchange of energy due to a pressure difference or other mechanical change in the surroundings. Heat is an exchange of energy due to a difference in temperature between the system and its surroundings.

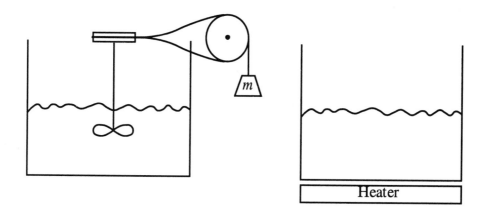

Critical Thinking Questions

7. Model 2 shows two beakers containing liquid water. Describe two processes that could alter the energy, ΔU_{sys}, of the water resulting in the same final temperature.

8. Based on Model 2, write an expression relating dU_{sys} to dq and dw for an infinitesimal change in the energy of the system. Pay close attention to the signs in your equation.

Information

There are a number of different circumstances under which chemical systems can undergo changes. Examples of several of these are:

- Isothermal processes

- Reversible processes

- Processes occurring at constant volume

- Processes occurring at constant pressure

- Processes for which no heat is exchanged with the surroundings. These are known as **adiabatic** processes.

Chemists find it useful to be able to calculate work, heat, and changes in energy of the system for these types of processes.

Critical Thinking Questions

9. What is dT for an isothermal process?

10. Consider an adiabatic, reversible process:

   a)   What is dq for an adiabatic process?

   b)   For a reversible process, how is dw related to pressure and volume?

   c)   Use your answer to CTQ 8 and parts a and b above to provide an equation relating dU_{sys} to pressure and volume for an adiabatic, reversible process.

11. For a constant volume process:

   a)   What is dw?

b)   Use your answer to CTQ 8 and CTQ 11a to provide an equation relating dU_{sys} to dq and dw for a constant volume process.

12.   For a constant pressure process, derive an expression to show how dq is related to dU_{sys}, P_{ex}, and dV.

Information

Energy is one of a class of thermodynamic functions known as exact differentials or state functions. This means that the path taken to get from one state of the system to another is irrelevant to the change in energy produced in this process. In other words, the change in a state function is independent of path and depends only on initial and final states. Neither q nor w is an exact differential. Thus, the heat or work associated with a process depends on the path taken to produce a change in state.

A system is not thought of as possessing certain amounts of work or heat at the beginning of a process and certain amounts at the end. Therefore, the symbol Δ (which usually means final minus initial in thermodynamics) is never used with q or w. By definition,

$$\int dw = w \qquad \int dq = q$$

Critical Thinking Questions

13.   What is the energy change for a system that undergoes a cyclic process which begins and ends in the same state?

14.   Describe two processes with different values for w that have the same initial and final states.

15.   Describe a process for which $\Delta U_{tot} = 0$.

Exercises

1. The work done by the system during a certain adiabatic process is 200 J. What is ΔU_{sys}?

2. An ideal gas is heated from 1 atm and 298 K to 2 atm and 398 K. The gas is then cooled to 1 atm and 298 K. What is ΔU_{sys}?

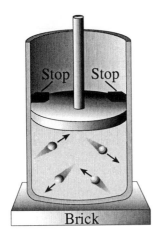

A piston and cylinder apparatus containing an ideal gas. The piston is held in place by stops.

3. A hot brick is placed in contact with a fixed container with an ideal gas. Consider the brick to be the system and the gas to be the surroundings. Initally, $U_{tot}^{init} = U_{sys}^{init} + U_{surr}^{init}$. The brick and the gas then exchange energy to reach a final state.

   a) Provide an expression for the final total energy, U_{tot}^{fin}.

   b) Provide an expression for ΔU_{tot} for the process in terms of the energies of the system and the surroundings.

   c) What is the magnitude of ΔU_{tot}?

4. a) Based on the first law of thermodynamics, what is the value of ΔU_{tot} for any process?

   b) Based on the first law of thermodynamics, provide an equation relating dU_{sys} to dU_{surr} and dU_{wt}.

   c) If an amount of heat dq is transferred from the surroundings to the system, what is dU_{surr} (in terms of dq)?

d)   When the heat dq is transferred to the system, the expansion of the system results in work dw being done by the system. In this case, $dw = -dU_{wt}$.

   i)   Use this result, in conjunction with your answers to b) and c) to derive an expression relating dU_{sys} to dq and dw.

   ii)   Compare your answer to d) i) with your answer to CTQ 8. Comment on the comparison.

5.   Is your answer to the focus question consistent with the first law of thermodynamics? Explain. Will all of the energy exchange between the brick and the water go into producing only thermal changes?

6.   If the same amount of heat is supplied to the following two containers will the temperature rise in the two containers be the same? Explain.

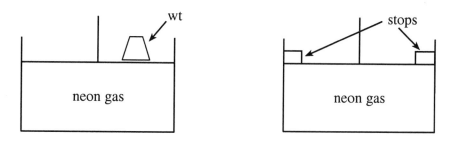

7.   If an ideal gas expands adiabatically against a constant external pressure, does the energy of the gas change? If so, how? Explain.

8.   If heat is added to a gas at constant volume, does the energy of the gas change? If so, how? Explain.

ChemActivity T3[*]

Enthalpy

Focus Question: If the same amount of heat is supplied to the following two containers, will the energy change of the gas in each container be the same? Explain.

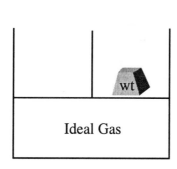 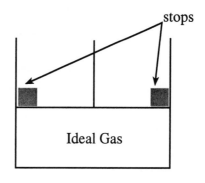

---

[*]Some material in this activity taken from J.N. Spencer, G.M. Bodner, and L.H. Rickard, *Chemistry: Structure and Dynamics*, John Wiley & Sons, Inc., 2nd Ed., 2002, Chpt. 7. This material is used by permission of John Wiley & Sons, Inc.

Model 1: An Ideal Gas in a Piston-Cylinder.

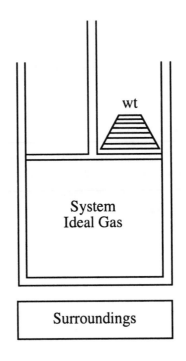

Enthalpy, H, is defined as:

$$H \equiv U + PV \tag{1}$$

The property enthalpy is defined this way because it is a convenient quantity in many situations, as will be seen shortly.

Critical Thinking Questions

1. Recall that energy, U, is a state function (or that dU is an exact differential).

   a) Is PV a state function?

   b) Is H a state function?

2. Use the definition of enthalpy, equation (1), to provide a general expression for dH.

3. Use your answer to CTQ 2 to provide an expression for dH for a constant pressure process.

4. Recall from CA 1 that $dw = -P_{ex} dV$. Recall how dU is related to dq and dw. Use your answer to CTQ 3 to provide an expression showing the relationship between dH and dq for a constant pressure process.

5. From your answers to CTQs 3 and 4 you can obtain equations involving ΔH for constant pressure processes. Since H is a state function, $\int_{H_1}^{H_2} dH = \Delta H$.

    a) Integrate both sides of your answer to CTQ 3 to obtain an equation for ΔH in terms of ΔU, $P_{ex} = P$, and ΔV for a constant pressure process.

    b) Integrate both sides of your answer to CTQ 4 to obtain a relationship between ΔH and q for a constant pressure process.

Model 2: Enthalpy as a State Function.

$$H_2(g) + Cl_2(g) \longrightarrow 2\ HCl(g) \qquad\qquad (1)$$

Because enthalpy is a state function we can visualize this reaction as occurring by two simple processes. First, we break the bonds in the starting materials to form hydrogen and chlorine atoms in the gas phase.

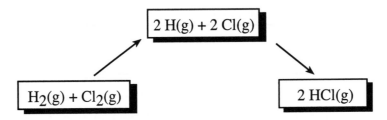

Then these atoms are recombined to form the product of the reaction.

Critical Thinking Question

6. Does the chemical transformation of H_2 and Cl_2 actually proceed by the steps given in Model 2?

Information

The symbol "Δ" usually refers to properties of products minus properties of reactants. Thus ΔH when used in conjunction with a chemical reaction means enthalpies of products minus enthalpies of reactants. Because the symbol ΔH can be ambiguous in some cases, the symbol $\Delta_r H$ is used in these worksheets whenever a chemical reaction or a change in phase occurs.

In order to break the covalent bonds in a mole of Cl_2 molecules 243.4 kJ of enthalpy are required.

$$Cl_2(g) \longrightarrow 2\ Cl(g) \qquad \Delta_r H = 243.4\ kJ/mol$$

It takes 435.3 kJ of enthalpy to break apart a mole of H_2 molecules to form hydrogen atoms.

$$H_2(g) \longrightarrow 2\ H(g) \qquad \Delta_r H = 435.3\ kJ/mol$$

Therefore the bond breaking procedure costs:

Bond breaking	$Cl_2(g) \longrightarrow 2\ Cl(g)$	$\Delta_r H = 243.4\ kJ/mol$
	$H_2(g) \longrightarrow 2\ H(g)$	$\Delta_r H = 435.3\ kJ/mol$
	$H_2(g) + Cl_2(g) \longrightarrow 2\ H(g) + 2\ Cl(g)$	$\Delta_r H = 678.7\ kJ/mol$

Now combine these atoms to make bonds and release heat.

Bond making $2 H(g) + 2 Cl(g) \longrightarrow 2 HCl(g)$ $\Delta_rH = -863.2$ kJ/mol

The overall change in the enthalpy of the system that occurs during this reaction can be calculated by combining Δ_rH for the two hypothetical steps in this reaction.

$$
\begin{array}{ll}
678.7 \text{ kJ/mol} & \text{Bond breaking} \\
\underline{-863.2 \text{ kJ/mol}} & \text{Bond making} \\
-184.5 \text{ kJ/mol} &
\end{array}
$$

According to this calculation, the overall reaction is exothermic (Δ_rH is negative) by a total of -184.5 kJ when one mole of hydrogen reacts with one mole of chlorine to form two moles of HCl.

$$H_2(g) + Cl_2(g) \longrightarrow 2 HCl(g) \qquad \Delta_rH = -184.5 \text{ kJ/mol} \qquad (2)$$

Critical Thinking Questions

7. In terms of the value of the Δ_rH for this reaction, does it matter whether the chemical transformation of H_2 and Cl_2 actually proceeds by the steps given in Model 2? Why or why not?

Information

At the heart of the reaction thermodynamics is the concept that the enthalpy of a system is a state function. As a result, *the value of Δ_rH for a reaction doesn't depend on the path used to convert the starting materials into the products of the reaction.* It only depends on the initial and final conditions—the reactants and products of the reaction.

Note that the units on Δ_rH in (2) above are kJ/mol, or more correctly, are kJ/mol$_{rxn}$ where mol$_{rxn}$ represents the chemical equation as an entire unit.

To have meaning, a change in enthalpy must be associated with a specific chemical equation. Although the term "mol$_{rxn}$" is used to describe the reaction, it does not necessarily mean that there is only 1 mole of reactant or product. In the above chemical equation there are 2 mol of reactants and 2 mol of product, but the reaction as a whole is referred to as one unit or one mole of chemical reaction.

Coefficients in a chemical equation are unitless but may be thought of as giving the number of moles of a substance per mole of reaction, for example $\dfrac{2 \text{ mol HCl}}{\text{mol}_{rxn}}$.

Model 3: Hess' Law.

Hess' law states that the enthalpy of reaction ($\Delta_r H$) is the same regardless of whether a reaction occurs in one step or in several steps. Thus the enthalpy change for a reaction can be calculated by adding the enthalpies associated with a series of hypothetical steps into which the reaction can be broken.

Information

The heat given off under standard-state conditions when water is formed from its elements as both a liquid and a gas has been measured. The standard state of a liquid or solid substance is the pure substance at 1 bar. The standard state for gases is the ideal gas at 1 bar. The superscript o indicates that all species are in their standard states. It is common practice to tabulate data at 25°C.

$$H_2(g) + \frac{1}{2} O_2(g) \longrightarrow H_2O(\ell) \qquad \Delta_r H° = -285.83 \text{ kJ/mol}$$

$$H_2(g) + \frac{1}{2} O_2(g) \longrightarrow H_2O(g) \qquad \Delta_r H° = -241.82 \text{ kJ/mol}$$

Critical Thinking Questions

8. For the reactions above what is the significance of the sign of $\Delta_r H°$?

9. What would be the enthalpy change for the reverse of the first reaction?

10. Use these data and Hess' law to calculate $\Delta_r H°$ for the following reaction.

$$H_2O(\ell) \longrightarrow H_2O(g)$$

Exercise

1. Before pipelines were built to deliver natural gas, individual towns and cities contained plants that produced a fuel known as "town gas" by passing steam over red-hot charcoal.

$$C(s) + H_2O(g) \longrightarrow CO(g) + H_2(g)$$

Calculate Δ_rH for this reaction from the following information.

$$C(s) + \frac{1}{2}O_2(g) \longrightarrow CO(g) \qquad \Delta_rH^\circ = -110.53 \text{ kJ/mol}$$

$$C(s) + O_2(g) \longrightarrow CO_2(g) \qquad \Delta_rH^\circ = -393.51 \text{ kJ/mol}$$

$$CO(g) + \frac{1}{2}O_2(g) \longrightarrow CO_2(g) \qquad \Delta_rH^\circ = -282.98 \text{ kJ/mol}$$

$$H_2(g) + \frac{1}{2}O_2(g) \longrightarrow H_2O(g) \qquad \Delta_rH^\circ = -241.82 \text{ kJ/mol}$$

Information

The enthalpy change for a reaction that produces one mole of a compound from its elements, the elements being in their stable states of aggregation at 1 bar (100000 Pa) and usually 298 K, is called the **enthalpy of formation**, $\Delta_rH^\circ_f$, for that compound. The standard state pressure was 1 atm (101325 Pa) for many years. Consequently most tables are tabulated using 1 atm rather than 1 bar. The difference, however, is slight and in this course no distinction will be made between bar and atmosphere.

Model 4: Some Chemical Reactions at 298 K.

a) $Mg(s) + CO(g) + O_2(g) \longrightarrow MgCO_3(s)$

b) $MgO(s) + CO_2(g) \longrightarrow MgCO_3(s)$

c) $Mg(s) + C(s) + \frac{3}{2}O_2(g) \longrightarrow MgCO_3(s)$

d) $BaCO_3(s) \longrightarrow BaO(s) + CO_2(g)$

e) $CO(g) + \frac{1}{2}O_2(g) \longrightarrow CO_2(g)$

f) $C(s) + O_2(g) \longrightarrow CO_2(g)$

Each gas in all of the reactions above is at one bar.

Critical Thinking Questions

11.   In which of the above reactions is one mole of compound produced?

12. In which of the above reactions is the product the result of the reaction of the elements that compose it, each of the elements being in their stable states at 1 bar and 298 K?

13. Which of the above reactions are enthalpy of formation reactions?

14. What does the symbol $\Delta_r H_f^o$ represent?

15. Calculate $\Delta_r H^\circ$ for the reaction

$$MgO(s) + CO_2(g) \longrightarrow MgCO_3(s)$$

from the following enthalpy of formation data.

$$Mg(s) + \frac{1}{2}O_2(g) \longrightarrow MgO(s) \qquad \Delta_r H_f^o = -601.70 \text{ kJ/mol}$$

$$C(s) + O_2(g) \longrightarrow CO_2(g) \qquad \Delta_r H_f^o = -393.51 \text{ kJ/mol}$$

$$Mg(s) + C(s) + \frac{3}{2}O_2(g) \longrightarrow MgCO_3(s) \qquad \Delta_r H_f^o = -1095.8 \text{ kJ/mol}$$

16. Is there a generalization of the use of enthalpy of formation data to calculate $\Delta_r H^\circ$ for a chemical reaction? If so what is it?

Exercises

2. Calculate $\Delta_r H^{\circ}_{298}$ for

$$Fe_2O_3(s) + 3\,H_2(g) \rightleftharpoons 2\,Fe(s) + 3\,H_2O(\ell)$$

$$\Delta_r H^{\circ}_{f_{298}} \;(kJ\,mol^{-1})$$

$Fe_2O_3(s)$	-824.2
$H_2O(\ell)$	-285.830

3. According to the criteria for an enthalpy of formation reaction, what must be the enthalpy of formation of any element in its stable state of aggregation at 1 bar?

4. If $\Delta_r H^{\circ}$ for the following reaction is -36.40 kJ mol^{-1}, what is the enthalpy of formation of HBr(g)?

$$1/2\,H_2(g) + 1/2\,Br_2(\ell) \rightleftharpoons HBr(g)$$

5. If $\Delta_r H^{\circ}_f$ for NH$_4$Cl(s) is -314.43 kJmol^{-1} what is $\Delta_r H^{\circ}$ for this reaction?

$$1/2\,N_2(g) + 2\,H_2(g) + 1/2\,Cl_2(g) \rightleftharpoons NH_4Cl(s)$$

6. $\Delta_r H^{\circ}$ for

$$N_2(g) + 3H_2(g) \rightleftharpoons 2NH_3(g)$$

is -92.2 kJ/mol (-92.2 kJ/mol$_{rxn}$).

   a)   What will be the enthalpy change if 0.200 mole of N$_2$ reacts?

   b)   How many moles of H$_2$ must be consumed if 0.200 mole of N$_2$ reacts?

   c)   How many moles of NH$_3$ can be formed if 0.200 mole of N$_2$ reacts?

   d)   Calculate $\Delta_r H^{\circ}$ for the reaction in b) and c).

   Pay careful attention to units and comment on the three values for $\Delta_r H^{\circ}$ calculated.

ChemActivity T3A

Enthalpy

Focus Question: Ethane and oxygen gas are put into a constant temperature piston-cylinder arrangement as shown below. The following reaction occurs:

$2C_2H_6(g) + 7O_2(g) \rightleftharpoons 4CO_2(g) + 6H_2O(g)$.

Will the weight be raised, lowered, or stay at the same height? Why?

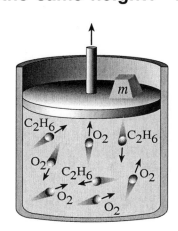

Information

When dealing with chemical reactions for which the numbers of moles are changing, it is important to remember that parameters describing the reaction will all have a unit referring to a mole of reaction, mol_{rxn}, even though the subscript rxn is not commonly used. Thus, as discussed previously, the units for the enthalpy change $\Delta_r H$ will be kJ/mol (kJ/mol_{rxn}). Units on other parameters that describe a change will correspondingly have the per mol_{rxn} unit. Thus for the change in volume that may be produced as the result of a

chemical reaction, the units on $\Delta_r V$ will be $\dfrac{volume}{mol}$ $\left(\dfrac{volume}{mol_{rxn}}\right)$.

Model 1: System is Expanded (or Contracted) by an External Source of Heat (or Cooling).

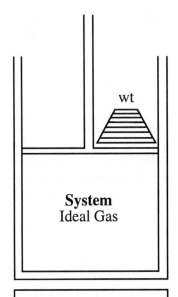

System
Ideal Gas

Surroundings:
initial—a hot brick;
final—a brick at the
same temperature as
the final temperature
of the gas.

The brick loses energy to the gas-cylinder. In this case the energy lost by the surroundings appears in the increased temperature of the gas and the increased height of the piston and weight. Not all of the energy lost by the surroundings has gone into increasing the temperature of the gas molecules. Some of the energy went into increasing the temperature of the gaseous molecules but some also went into the work of lifting the piston and weight.

When a symbol does not have a subscript sys or surr attached, it is understood that the thermodynamic parameters are those of the system.

Critical Thinking Questions

1. Is the process described in Model 1 a constant pressure process?

2. Is this a constant temperature process?

3.   Is q for this process zero? Negative? Positive?

4.   Recall that for a constant pressure process that $\Delta H = q_P$. Is ΔH for this process zero? Negative? Positive?

Exercise

1.   Suppose that the brick in Model 1 loses 500 J and the volume change of the system is 1.43 L . If the pressure is 1.00 bar, what is q? ΔH? w? ΔU?

Model 2:   System is Expanded (or Contracted) by an Internal Chemical Reaction at Constant Temperature.

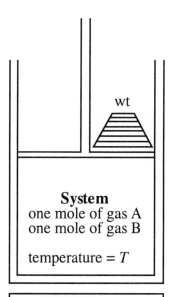

wt

System
one mole of gas A
one mole of gas B

temperature = T

Surroundings:
A bath that is so large
that its temperature, T,
remains constant
regardless of heat flow
in or out.

The chemical reaction of A and B goes to completion:

$$A(g) + B(g) \rightleftharpoons C(g) + 2\,D(g) \qquad \Delta_r H = 0$$

Critical Thinking Questions

5. Is the process described in Model 2 a constant pressure process?

6. Is it a constant temperature process?

7. Before the chemical reaction occurs, what is the total number of gaseous moles in the container?

8. After the chemical reaction occurs, what is the total number of gaseous moles in the container?

9. What is $\Delta_r n_{gaseous}$ for this process?

10. What is the expression for $\Delta_r(PV)$ for this process (assume that all gases are ideal)?

11. What happens to the position of the piston?

12. Is w for this process zero? Negative? Positive?

13. Suppose the chemical reaction is:

$$A(g) + B(g) \rightleftharpoons C(g) + 2\,D(g) \qquad \Delta_r H < 0$$

If the reaction goes to completion, does the piston move more, less, or the same as it did originally in Model 2? Explain.

14. Suppose the chemical reaction is:

$$A(g) + B(g) \rightleftharpoons C(g) + D(g) \qquad \Delta_r H < 0$$

If the reaction goes to completion, does the piston move more, less, or the same as it did originally in Model 2? Explain.

15. Suppose the chemical reaction is:

$$A(g) + 2\,B(g) \rightleftharpoons C(g) + D(g) \qquad \Delta_r H < 0$$

Assuming that the reaction goes to completion, describe how the piston moves in this case. Explain.

16. For a reaction involving ideal gases at constant temperature, what determines the amount of work which can be done? Explain clearly.

Exercise

2. Suppose that in a system such as in Model 2, one mole of $A(g)$ and one mole of $B(g)$ react to form one mole of $C(g)$ and two moles of $D(g)$. If the temperature is 298 K and the pressure is 1.0 bar, calculate q, $\Delta_r n$, $\Delta_r(PV)$, w, and $\Delta_r U$ when

a) $\Delta_r H = 0$

b) $\Delta_r H < 0$

If it is not possible to calculate the magnitude of any quantity, indicate whether it is positive, negative, or zero.

Model 3: System is Expanded (or Contracted) by an Internal Chemical Reaction Adiabatically.

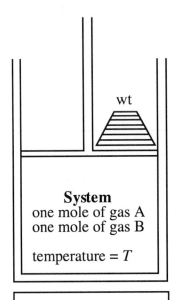

The chemical reaction of A and B is as follows:

$$A(g) + B(g) \;\rightleftharpoons\; C(g) + D(g) \qquad \Delta_r H < 0$$

The temperature in the container increases not as a result of externally supplied energy but because of the energy produced by the breaking and making of chemical bonds that occur in the reaction. The weight rises and work is done. The heat for this process came not from an external thermal source but from a chemical process. Heat was evolved by the chemical reaction because the sum of bond strengths of the products is greater than that of the reactants. Part of the evolved heat went into the raising of the piston and part into heating the container and gases. Under these conditions of constant pressure the heat is referred to as the enthalpy change produced by the chemical reaction.

Critical Thinking Questions

17. Is this a constant pressure process?

18. Is this a constant temperature process?

19. When the chemical reaction occurs, what happens to the temperature of the system?

20. Is this an adiabatic or an isothermal process?

21. What happens to the position of the piston?

22. Is w for this process zero? Negative? Positive?

23. When chemical reactions occur, give two sources for any work that may be done.

Exercises

3. Suppose that in Model 2, one mole of A(g) and one mole of B(g) react (let $\Delta_r H = -1.00$ kJ/mole). If the volume change is 2.50 L and the pressure is one bar, calculate q, w, and $\Delta_r U$.

4. Suppose that in Model 3, one mole of A(g) and one mole of B(g) react (let $\Delta_r H = 0$). If the temperature is 298 K and the pressure is one bar, calculate q, w, and $\Delta_r U$.

5. For a chemical reaction at constant T and P show how $\Delta_r U^\circ$ is related to $\Delta_r H^\circ$.

6. For the constant volume combustion of one mole of benzene

$$C_6H_6(\ell) + \frac{15}{2} O_2(g) \rightleftharpoons 6\, CO_2(g) + 3\, H_2O(\ell)$$

3623 kJ of heat were evolved at 298 K. What is $\Delta_r U^\circ$ for this process? What would be $\Delta_r H^\circ$ at 298 K and 1 bar for this reaction?

7. What different meanings do the two symbols $\Delta_r H°$ and $\Delta H°$ convey?

8. Suppose the process described in Model 2 is carried out. How could you calculate w for the process from only a knowledge of the reaction stoichiometry and P_{ex}?

Heat Capacity

Focus Question: **Equal masses of water at 50 °C and Hg at 80° C are mixed. What is the temperature of the resulting mixture?**

a) **less than 65 °C**
b) **65 °C**
c) **more than 65 °C**
d) **unable to determine without knowing densities**

Model 1: Temperature, Heat, and Energy.

Temperature is a measure of the molecular motion of particles. Heat can be used to cause a change in temperature. The higher the temperature the greater the molecular motion. Thermal energy (heat) can be absorbed to increase the kinetic energy of motion of the particles composing a solid, liquid, or gas. The energy absorbed can also weaken the interactions among molecules causing them to move farther apart. In this case, the energy absorbed has increased the potential energy. Because the structure and intermolecular interactions of compounds vary greatly, different amounts of heat are necessary to raise the temperature by a given amount.

Critical Thinking Questions

1. If energy is absorbed to cause an increase in kinetic energy of translation of particles, does the temperature of the system increase?

2. Consider 1 mole samples of Ne and N_2 at the same temperature T. Equal amounts of heat are added to each sample under otherwise identical conditions.

   Predict whether the final temperatures of the two samples will be the same or different. If different, predict which will have the higher final temperature. Explain clearly.

Table 1. Temperature Rise Observed When 20 J of Energy is Transferred to 1 mole of Various Substances at 25 °C and 1 bar

Substance	Temperature Rise (°C)
Ne (g)	1
N_2(g)	0.7
CH_4 (g)	0.6
H_2O (l)	0.3

The amount of heat required to raise the temperature of different systems by the same amount is called the **heat capacity, C**. The **molar heat capacity**, with units of J mol^{-1} K^{-1}, is the heat required to raise the temperature of *one mole* of a substance by one degree and is indicated by a super bar over the heat capacity symbol $\overline{C}$.

Critical Thinking Questions

3   Based on the data in Table 1, what is the molar heat capacity of Ne(g)?

4. Based on the data presented, rank the 4 substances in Table 1 from highest to lowest molar heat capacity. Explain your reasoning.

5. Predict whether the molar heat capacity for H_2O(g) at 25° C and 1 bar would be greater than, less than, or about the same as the molar heat capacity for $H_2O(\ell)$ as shown in Table 1. Explain your reasoning.

Model 2: Heat Capacity.

Mathematically, heat capacity, C, is defined by

$$C = \frac{dq}{dT} \tag{1}$$

Heat and work are not exact differentials and hence their magnitudes depend on the path to determine them. Thus, heat capacity will differ depending on the conditions under which it is measured.

The heat capacity determined under conditions of constant volume is

$$C_V = \frac{dq_V}{dT} = \left.\frac{\partial U}{\partial T}\right)_V \tag{2}$$

The heat capacity determined under conditions of constant pressure is

$$C_P = \frac{dq_P}{dT} = \left.\frac{\partial H}{\partial T}\right)_P \tag{3}$$

Table 2. Examples of the Molar Heat Capacities $\overline{C}_V$ and $\overline{C}_p$ at 25 °C

Substance	$\overline{C}_V$ (J K^{-1} mol^{-1})	$\overline{C}_P$ (J K^{-1} mol^{-1})
Gases:		
Helium, He	12.47	20.79
Neon, Ne	12.47	20.79
Nitrogen, N_2	20.81	29.12
Methane, CH_4	27.00	35.31
Sulfur hexafluoride, SF_6	88.97	97.28
Liquids:		
Water, H_2O	74.66	75.29
Benzene, C_6H_6	92.9	135.6
Solids:		
Diamond, C	6.11	6.11
Quartz, SiO_2	44.12	44.43
Sodium chloride, NaCl	47.7	50.50

Critical Thinking Questions

6. Energy, U, may be considered a function of T and V: $U = U(T,V)$.

    a)   Write the total differential for $U = U(T,V)$.

    b)   Substitute the molar heat capacity, $\overline{C_V}$, for the appropriate partial derivative.

    c)   Obtain an expression for dU in terms of $\overline{C_V}$ for a constant volume process.

7. Consider a process for which $P_{ex} = P_{int}$ at all times.

    a)   Write an expression for dw under these conditions.

    b)   Substitute your expression for dw from part a) above into the first law.

    c)   Use the expression from part b) to obtain a relationship between dU and dq_V under constant volume conditions.

    d)   Use the expression from part b) to obtain a relationship between dU and dq_P under constant pressure conditions.

    e)   Provide an expression relating dq_P to dH.

f) Based on your answer to part e) above (and equation 3), provide an expression relating dH to the molar heat capacity, $\overline{C}_P$, for a constant pressure process.

8. Enthalpy, H, can be considered a function of T and P: $H = H(T, P)$

 a) Write the total differential for $H = H(T, P)$.

 b) Use the expression in part a) and make the appropriate substitutions to obtain an expression for dH in terms of C_P for a constant pressure process. How does this expression compare to the one you obtained in 7f above?

9. Describe a process that could be used to determine $\overline{C}_V$.

10. Describe a process that could be used to determine $\overline{C}_P$.

Model 3: Ideal Gases.

An ideal gas has two distinguishing characteristics:

- An ideal gas obeys the equation of state $PV = nRT$.

- The energy of an ideal gas is a function of temperature only.

Critical Thinking Questions

11. Based on the information in Model 3, what is $\left.\dfrac{\partial U}{\partial P}\right)_T$? $\left.\dfrac{\partial U}{\partial V}\right)_T$?

12. Use the total differential for $U(T, V)$ from CTQ 6 to obtain an expression for dU for an ideal gas.

13. Calculate ΔU for an isothermal process for an ideal gas in which the pressure increases from 1 bar to 10 bar.

Exercises

1. In the constant volume addition of heat to the system

$$\frac{dq_V}{dT} = \left.\frac{\partial U}{\partial T}\right)_V = C_V$$

In the constant pressure process

$$\frac{dq_P}{dT} = \left.\frac{\partial H}{\partial T}\right)_P = C_P$$

Show that for an ideal gas $\overline{C_P} - \overline{C_V} = R$. Hint: How much more heat is required to raise the temperature of the gas by 1 K if the process is carried out at constant pressure rather than constant volume? Does $\overline{C_P} - \overline{C_V} = R$ agree with the data given in Table 2? Explain.

2. How much heat is required to raise the temperature of one mole of NaCl(s) by 15°C at constant pressure? Assume that the heat capacity does not change with temperature and use Table 2.

Model 4: Heat Capacities as a Function of Temperature.

Heat capacities vary with temperature. The variation with temperature observed experimentally for a variety of substances follows the same general functional form:

$$\overline{C_P} = a + bT + cT^{-2} \qquad\qquad (4)$$

where a, b, and c are constants, as shown in Table 3.

Table 3. Molar Heat Capacities in $JK^{-1}mol^{-1}$ at Constant Pressure (Parameters for the equation $\overline{C}_P^{\,\circ} = a + bT + cT^{-2}$.)

	a	$b \times 10^3$	$c \times 10^{-5}$
Gases (in Temperature Range 298 to 2000 K)			
He, Ne, Ar, Kr, Xe	20.79	0	0
S	22.01	−0.42	1.51
H_2	27.28	3.26	0.50
O_2	29.96	4.18	−1.67
N_2	28.58	3.76	−0.50
S_2	36.48	0.67	−3.76
CO	28.41	4.10	−0.46
F_2	34.56	2.51	−3.51
Cl_2	37.03	0.67	−2.84
Br_2	37.32	0.50	−1.25
I_2	37.40	0.59	−0.71
CO_2	44.22	8.79	−8.62
H_2O	30.54	10.29	0
H_2S	32.68	12.38	−1.92
NH_3	29.75	25.10	−1.55
CH_4	23.64	47.86	−1.92
TeF_6	148.66	6.78	−29.29
Liquids (from Melting Point to Boiling Point)			
I_2	80.33	0	0
H_2O	75.48	0	0
NaCl	66.9	0	0
$C_{10}H_8$	79.5	407.5	0

Solids (from 298 K to Melting Point, or 2000 K)			
C(graphite)	16.86	4.77	−8.54
Al	20.67	12.38	0
Cu	22.63	6.28	0
Pb	22.13	11.72	0.96
I_2	40.12	49.79	0
NaCl	45.94	16.32	0
$C_{10}H_8$	−115.9	937	0

Source: Calculated from data of G.N. Lewis and M. Randall, *Thermodynamics*, 2d ed. (rev. by K.S. Pitzer and L. Brewer), McGraw-Hill, 1961, with permission of the McGraw-Hill Companies.

Critical Thinking Questions

14. Show that

    a)   for a constant pressure process: $\Delta H = \displaystyle\int_{T_1}^{T_2} \overline{C_P}\, dT$

    b)   for a constant volume process: $\Delta U = \displaystyle\int_{T_1}^{T_2} \overline{C_V}\, dT$

    c)   Under what circumstances is $\Delta H = \overline{C_P}\,(T_2 - T_1);\ \Delta U = \overline{C_V}\,(T_2 - T_1)$?

d)   Use equation 4 for $\overline{C}_P$ to find ΔH as a function of temperature for a constant pressure process.

Exercises

3.   Calculate the heat required to raise one mole of $CO_2(g)$ from 0°C to 300°C

a)   at constant pressure

b)   at constant volume

$$\overline{C}_P = 44.22 + 8.79 \times 10^{-3}\,T - 8.62 \times 10^5\,T^{-2}\ \text{JK}^{-1}\text{mol}^{-1}$$

$$\overline{C}_V = 35.91 + 8.79 \times 10^{-3}\,T - 8.62 \times 10^5\,T^{-2}\ \text{JK}^{-1}\text{mol}^{-1}$$

4.   How much heat is required to raise the temperature of one mole of NaCl(s) from 25°C to 40°C at constant pressure? Do not assume that the heat capacity is independent of temperature and compare your answer to Exercise 2.

5.   Consider an ideal gas undergoing an isothermal process. Our goal is to determine ΔH for this situation.

a)   Write the total differential for $H = H(T,P)$ and substitute C_P for the appropriate partial derivative.

b)   Use the definition of H in terms of U and PV to evaluate $\left(\dfrac{\partial H}{\partial P}\right)_T$. Remember that an ideal gas is involved and that the process is isothermal.

c)   Calculate ΔH from the total differential in part a) making appropriate substitutions.

6.   If the energy of an ideal gas is a function of temperature only, what conclusions can be made concerning the enthalpy of an ideal gas?

Table 4. $\overline{C}_P^{\,\circ}$ as a Function of Temperature

	$\overline{C}_P^{\,\circ}$ (JK^{-1}mol^{-1})	T(K)
H$_2$O(g)	33.59	298
	35.23	500
	41.27	1000
C(graphite)	8.52	298
	14.62	500
	21.61	1000
H(g)	20.79	298
	20.79	500
	20.79	1000
H$_2$(g)	28.84	298
	29.26	500
	30.21	1000

R.A. Alberty and R.J. Silbey, *Physical Chemistry*, John Wiley & Sons, Inc. NY, 1992. This material is used by permission of John Wiley & Sons, Inc.

Critical Thinking Questions

15. Critique the following statement: The more complex the species the larger is $\overline{C}_P^{\,\circ}$ and the larger the increase in $\overline{C}_P^{\,\circ}$ with increasing temperature. Refer to Table 4.

16. Use your knowledge of the temperature dependence of $\overline{C}_P^{\,\circ}$ to explain why $\Delta_r H^\circ$ varies with temperature.

ChemActivity T5

Temperature Dependence of the Enthalpy of Reaction

Focus Question: When water evaporates at 298 K, will the enthalpy change be the same as at 373 K? Will the enthalpy change when $O_2(g)$ and $H_2(g)$ react to form $H_2O(g)$ be the same at 298 K and 373 K?

Model: A Reaction at Different Temperatures.

Values for $\Delta_r H$ for a chemical reaction are often tabulated at a given temperature - for example, 298 K. It is often desirable to know $\Delta_r H$ at another temperature.

At T_2 :   Reactants → Products $\Delta_r H_{T_2}$ (1)

At T_1 :   Reactants → Products $\Delta_r H_{T_1}$ (2)

For the above processes, P is constant.

Critical Thinking Questions

1. Is the enthalpy of the products changed when the temperature is changed from T_1 to T_2? That is, for the constant pressure process

$$\text{Products at } T_1 \rightarrow \text{ Products at } T_2$$

is the value of ΔH zero or nonzero? If zero, explain why. If nonzero, provide an expression for evaluating ΔH for this process.

2.   Recall that $\Delta_r H = H(\text{products}) - H(\text{reactants})$. Why might $\Delta_r H$ be different for a given chemical reaction at different temperatures?

3. Recall that $\Delta_r H$ is a state function. Construct a cycle using the model above which includes processes (1) and (2) above and also the change in temperature for reactants and products. From this cycle show how $\Delta_r H_{T_2}$ is related to $\Delta_r H_{T_1}$ and the heat capacities of reactants and products.

4. Generalize the expression obtained in CTQ 3 to show that $d\Delta_r H = \Delta_r \overline{C_P}\, dT$. Clearly define all terms.

5. Integrate the expression from CTQ 4 to obtain an expression for $\Delta_r H$ in terms of $\Delta_r \overline{C_P}$ under the assumption that $\Delta_r \overline{C_P}$ is not a function of temperature.

Exercises

1. Calculate $\Delta_r H°$ for the freezing of water at one bar and $-10°C$.

   The heat capacities are $H_2O(s) = 36.9 \ JK^{-1}mol^{-1}$ and $H_2O(\ell) = 74.5 \ JK^{-1} \ mol^{-1}$. $\Delta_r H°_{273}$ is $-6025 \ Jmol^{-1}$. Assume $\overline{C_P}$ is independent of temperature.

2. Use $\overline{C_P}$ data from ChemActivity 4 to find $\Delta_r H°$ at 500 K for

   $$H_2O(g) \rightleftharpoons H_2(g) + 1/2 \ O_2(g)$$

   Note that $\Delta_r H°$ at some temperature must be known.

3. Find $\Delta_r H°_{363}$ for

   $$C_6H_6(g) + 15/2 \ O_2(g) \rightleftharpoons 6 \ CO_2(g) + 3 \ H_2O(g)$$

   $$\Delta_r H^{Ovap}_{bz_{353}} = 30,800 \ J \ mol^{-1} \qquad \Delta_r H^{Ovap}_{H_2O_{373}} = 40,700 \ J \ mol^{-1}$$

   The normal boiling point of benzene is 353 K.

	$\overline{C_P^°}$ ($JK^{-1}mol^{-1}$)	
benzene (ℓ)	136	From $\Delta_r H_f°$ tables for reacton as written:
benzene (g)	82	$\Delta_r H°_{298} = -3169.4$
$O_2(g)$	30	
$CO_2(g)$	37	
$H_2O(l)$	75	
$H_2O(g)$	34	

4. Using data from Exercise 3 find the enthalpy of vaporization of water at 298 K and 1 bar pressure. State any assumptions.

ChemActivity T6

Entropy

Focus Question: **Indicate whether the following statement is true or false and explain your reasoning:**

When a hot brick is dropped into cold water, the temperature of the brick must decrease to be consistent with the first law of thermodynamics.

Model 1: The Second Law of Thermodynamics.

The direction in which a natural process occurs is summarized by the **second law of thermodynamics**, which states that natural processes are spontaneous when they lead to an increase in the entropy of the universe. The term *universe* means everything that might conceivably have its entropy altered as a result of the process.

Entropy, given the symbol S, is a measure of the number of ways energy is distributed or spread among particles. When energy, often in the form of heat, is supplied to a substance, that energy is stored by the translational, rotational, vibrational, and electronic energy levels and through intermolecular interactions. The greater the number of available energy storage modes a species has, the larger will be its entropy. Cyclopentene has a lower entropy than 1-pentene because the cyclopentene structure restricts the freedom of motion of its constituent atoms more than does the 1-pentene structure. The cyclopentene energy levels are more widely spaced and less available. When a gas is heated at a fixed volume, the added energy is shared among modes and spread in more ways than are possible in a colder gas, and the heated gas has a higher entropy.

The entropy change for a process is a measure of the removal or addition of constraints to the atoms, ions, or molecules that occurs during the process. The melting of a solid or the breaking apart of a larger molecule into smaller molecules allow an increase in the number of ways energy can be distributed and the entropy increases. In a chemical reaction, energy is distributed over the products and reactants according to the availability of their storage modes. The extent of the distribution of energy through the various modes of energy storage is at maximum when thermodynamic equilibrium is reached.

Entropy like energy and enthalpy is a state function; entropy is additive, but entropy is not conserved.

It is important to recognize that the second law of thermodynamics describes what happens to the entropy of the *universe*, not the system in which we are interested. The change in the entropy of the universe, or the total entropy change that accompanies a chemical reaction, is equal to the sum of the changes in the entropy of everything that might undergo an entropy change. Chemists divide these changes into what are called the *system* and *surroundings*.

$$\Delta S_{univ} = \Delta S_{sys} + \Delta S_{surr} = \Delta S_{tot}$$

Another way of stating the second law of thermodynamics is that for an event to occur spontaneously, ΔS_{univ} or ΔS_{tot} must be positive.

$$\Delta S_{univ} \geq 0$$

If the inequality applies, the process is irreversible; it will occur naturally. If the equality applies, the process is reversible or at equilibrium.

Chemists generally choose the chemical reaction as the system in which they are interested. The term *surroundings* is then used to describe the environment that is altered as a consequence of the reaction. Unless otherwise specified, the symbol ΔS will always refer to the system.

Critical Thinking Questions

1. Without reference to entropy, use a grammatically correct English sentence to describe what it means to indicate that a particular process is "spontaneous".

2. If $(S_{tot})_{final} > (S_{tot})_{initial}$:

    a)   what is the sign of ΔS_{tot}?

    b)   is the process considered to be spontaneous?

3. If $(S_{tot})_{final} < (S_{tot})_{initial}$:

    a)   what is the sign of ΔS_{tot}?

    b)   is the process considered to be spontaneous?

    c)   is the reverse process considered to be spontaneous?

4. If $(S_{tot})_{final} = (S_{tot})_{initial}$:

    a)   what is the sign of ΔS_{tot}?

    b)   is the process considered to be spontaneous?

    c)   is the reverse process considered to be spontaneous?

    d)   is the process at equilibrium?

5. Provide an expression for ΔS_{tot} in terms of the system and the surroundings.

6. Imagine tossing a hot brick into cold water in an adiabatic enclosure. Assume that the resulting process does not affect the volume of the brick or the water.

    a)   Define the system and the surroundings.

    b)   What will happen, in terms of temperature changes, after the brick is tossed into the water?

    c)   Is your answer to b) based on an application of the first law of thermodynamics? If so, explain how. If not, explain why not.

d)   What will be ΔU_{tot} for the process described in b)? Explain your reasoning.

e)   Is your answer to d) based on an application of the first law of thermodynamics? If so, explain how. If not, explain why not.

f)   Can you determine the sign and/or magnitude of ΔS_{tot} for the process described in b)? If so, provide this information. If not, explain why not.

7. Use a grammatically correct English sentence to describe the relationship between the concept of "equilibrium" and "reversible processes".

8. Use a grammatically correct English sentence to describe the relationship between "spontaneous" and "irreversible processes".

Model 2: Entropy Changes Under Various Conditions.

For a reversible process

$$dS = \frac{dq_{rev}}{T} \quad . \tag{1}$$

For any process at constant volume and temperature:

$$\Delta S_{surr} = \frac{\Delta U_{surr}}{T} \quad . \qquad\qquad T, V = \text{constant} \tag{2}$$

For any process at constant pressure and temperature:

$$\Delta S_{surr} = \frac{\Delta H_{surr}}{T} \quad . \qquad\qquad T, P = \text{constant} \tag{3}$$

Critical Thinking Questions

9. Consider any process which occurs at constant pressure and constant temperature. We are interested in determining the value of ΔS_{surr}. Consider a transfer of heat dq_{surr} to the surroundings.

    a) How is dq_{surr} related to dH_{surr} under these conditions?

    b) Is dH_{surr} an exact differential? What does this imply about dq_{surr}?

    c) Show how equation (3) can be obtained from equation (1).

10. A process occurs at constant pressure and temperature.

    a) Use the definition of ΔS_{tot} and equation (3) above to provide an expression relating ΔS_{tot}, ΔS_{sys}, and ΔH_{surr}.

    b) How is ΔH_{surr} related to ΔH_{sys} under these conditions? Explain.

    c) Show that $\Delta S_{tot} = \Delta S_{sys} - \dfrac{\Delta H_{sys}}{T}$. What are the restrictions on the generality of this expression? That is, what, if any, variables must be held constant? Is this expression restricted to reversible and/or irreversible processes? Explain.

11. For each case below, indicate whether ΔS_{tot} is greater than, less than, or equal to zero, or indicate that this can not be determined. Explain your reasoning.

  a)   reversible process (at constant T, P)

  b)   irreversible process (at constant T, P)

12. We can express dS in terms of energy, temperature, volume, and external pressure.

  a)   Starting with the first law of thermodynamics expressed in terms of dq and dw, use Model 2 to derive the following expressions:

$$dU = TdS - P_{ex}\,dV \qquad\qquad dS = \frac{dU}{T} + P_{ex}\frac{dV}{T}$$

  b)   Based on the expressions in a) above, if the temperature of an ideal gas is increased reversibly under conditions of constant external pressure, is the entropy change positive, negative, zero, or unable to be determined? Explain your reasoning.

Exercises

1.   Show how equation (2) can be derived from equation (1).

2.   For the process

$$H_2O(s) \rightleftharpoons H_2O(\ell)$$

at 1 bar and 0°C, $\Delta_r H_{273}^{\circ} = 6025$ Jmol^{-1}. What is $\Delta_r S_{273}^{\circ}$?

3.   For the combustion of glucose at 298 K, $\Delta_r S^{\circ} = 182.4$ JK^{-1}mol^{-1} and $\Delta_r H^{\circ} = -2808$ kJ mol^{-1}. Calculate ΔS_{tot}.

$$C_6H_{12}O_6(s) + 6\ O_2(g) \rightleftharpoons 6\ CO_2(g) + 6\ H_2O(\ell)$$

4 .   At one bar and 298 K, $\Delta_r S^{\circ}$ for

$$H_2(g) + 1/2\ O_2(g) \rightleftharpoons H_2O(\ell)$$

is -163.34 JK^{-1}mol^{-1} and $\Delta_r H^{\circ} = -285.83$ kJmol^{-1}. Is this reaction spontaneous? Does the sign of $\Delta_r S^{\circ}$ agree with your expectations?

5.   For C(graphite) = C(diamond) at 298 K and one bar, $\Delta_r H^{\circ} = +1.895$ kJmol^{-1} and $\Delta_r S^{\circ} = -3.363$ JK^{-1}mol^{-1}. Is this reaction spontaneous? Is the reverse reaction spontaneous?

6.   If $dU = C_V dT$ show that for the reversible expansion of one mole of an ideal gas:

$$\Delta S = C_V \ln \frac{T_2}{T_1} + R \ln \frac{V_2}{V_1}$$

if $C_V \neq f(T)$.

7.   Use the definition of enthalpy, the first law, Model 2, and that entropy is a state function to show that for the reversible expansion of one mole of an ideal gas:

$$\Delta S = C_P \ln \frac{T_2}{T_1} - R \ln \frac{P_2}{P_1}. \quad C_P \neq f(T).$$

8.   An ideal gas is expanded from 10 bar to 1.0 bar at constant temperature. Calculate ΔU, ΔH, and ΔS. $\overline{C_P} = 5/2\ R$.

9.   Calculate ΔS for

$$N_2\ (1\ \text{bar, } 298\ \text{K}) \longrightarrow N_2\ (32\ \text{bar, } 1500\ \text{K})$$

Assume N_2 to be an ideal gas.

$$\overline{C_P} = 28.58 + 3.77 \times 10^{-3}\ T - 0.50 \times 10^5\ T^{-2}\ \text{J K}^{-1}\ \text{mol}^{-1}$$

10.   The enthalpy of vaporization of benzene at 1 atm and its normal boiling point 353 K is 30.8 kJmol^{-1}. What is the difference in entropy between gaseous and liquid benzene under these conditions?

Entropy Changes as a Function of Temperature

Focus Question: **As temperature is increased, does the *difference* in entropy of gaseous and liquid water increase, decrease, or stay the same?**

Model 1: Entropy Changes.

S is a state function.

$$dS = \frac{dq_{rev}}{T}$$
(reversible process)   (1)

$$\Delta S_{sys} + \Delta S_{surr} = 0$$
(reversible process)   (2)

$$\Delta S_{sys} + \Delta S_{surr} > 0$$
(spontaneous process)   (3)

$$dU = TdS - PdV$$
(4)

Critical Thinking Questions

1. Show how equation (4) can be derived from the first law of thermodynamics for a reversible process.

2. Which of the variables in equation (4) is an exact differential (or state function)?

3. Explain why, although equation (4) was derived for a reversible process, it is applicable to all processes.

4. Let $dU = \overline{C_V}\,dT$ and rearrange equation (4) to provide an expression for dS for one mole of an ideal gas in terms of T, V, and C_V.

5. Write the total differential for $S = S(T,V)$.

6. Based on your answers to CTQs 4 and 5, find the derivatives

$$\left.\frac{\partial S}{\partial T}\right)_V \qquad \left.\frac{\partial S}{\partial V}\right)_T$$

for an ideal gas.

7. Derive an expression relating dS to $\overline{C_P}$, T, and P for an ideal gas. (Hint: Consider the definition of H and the first law.)

8. From the total differential of $S = S(T, P)$ and the results of CTQ 7:

    a) find $\left. \dfrac{\partial S}{\partial T} \right)_P$

    b) Is the expression obtained in part a) applicable only to an ideal gas, or is it generally valid? Explain your reasoning.

9. Use a grammatically correct English sentence to explain:

    a) the meaning of the symbol $\Delta_r S$.

    b) the meaning of the derivative $\left. \dfrac{\partial \Delta_r S}{\partial T} \right)_P$.

    c) how the derivative in part b) is related to the heat capacities of the reactants and the products.

10. For a constant pressure process for which $\Delta_r C_P$ is not a function of temperature, show that

$$\Delta_r S^{\circ}_{T_2} - \Delta_r S^{\circ}_{T_1} = \Delta_r C^{\circ}_P \ln \frac{T_2}{T_1}$$

11. Consider a constant pressure process in which $\Delta_r C_P$ is greater than zero and does not depend on temperature. If temperature is raised, does the value of $\Delta_r S$ increase, decrease, stay the same, or is it impossible to determine? Explain your reasoning.

Model 2: The Melting of Ice.

$$H_2O(s) \rightleftharpoons H_2O(l) \ . \qquad\qquad 273 \text{ K, 1 bar}$$

$$\Delta_r H^{\circ}_{273} = 6025 \text{ J mol}^{-1}$$

$$\Delta_r S^{\circ}_{273} = 22.1 \text{ J K}^{-1} \text{ mol}^{-1}$$

Critical Thinking Questions

12. Explain why both $\Delta_r H^{\circ}$ and $\Delta_r S^{\circ}$ are positive for the above phase transition.

13. Confirm that $\Delta_r S^{\circ} = \dfrac{\Delta_r H^{\circ}}{T}$ for the melting of ice at 273 K, 1 bar.

14. Complete the following table, assuming that $\Delta_r \overline{C}_P = 0$. (This assumption is not rigorously correct, but simplifies the calculation.)

Table 1.

	+10°C	0°C	−10°C
$\Delta_r H_T^{\circ}$			
ΔS_{surr}			
$\Delta_r S_{sys}$			
ΔS_{tot}			

15. For each temperature in the above table, explain what the value of ΔS_{tot} implies about the melting of ice.

16. Refer to CTQ's 14 and 15 to answer the following:

   a)   What is the relationship of dS to $\frac{dq}{T}$ for processes which cannot occur?

   b)   What is the relationship of dS to $\frac{dq}{T}$ for spontaneous processes?

   c)   What is the relationship of dS to $\frac{dq}{T}$ for equilibrium processes?

Exercises

1. a)   What is the entropy change if one mole of liquid water is heated from 0°C to 100°C under constant pressure? $\overline{C}_P^{\,\circ} = 75.5$ JK^{-1}mol^{-1} over this temperature interval.

    b)   Calculate $\Delta_r S$ for the transformation

$$H_2O \text{ (s, 0°C, 1 atm)} \longrightarrow H_2O \text{ (g, 100°C, 0.50 atm).}$$

$\Delta_r H_{fus_{273}} = 6025$ Jmol^{-1}, $\Delta_r H_{vap_{373}} = 40.7$ kJmol^{-1}

State all assumptions.

2. One mole of an ideal gas for which $\overline{C}_V = 3/2\,R$ is subjected to the following sequence of steps:

    a)   The gas is heated reversibly at a constant pressure of 1.00 bar from 25° to 100°C.

    b)   The gas is expanded isothermally and reversibly to double its volume.

    c)   The gas is cooled reversibly and adiabatically to 35°C.

    Calculate ΔU, ΔH, q, ΔS, and w for the overall process.

3. a)   Three moles of an ideal gas expand isothermally against a constant opposing pressure of 1bar from 6.7 to 67 liters. Calculate q, w, ΔU, ΔH, and ΔS.

    b)   Three moles of an ideal gas at 0°C expand isothermally and reversibly from 6.7 to 67 liters. Calculate q, w, ΔU, ΔH, and ΔS.

4. $\overline{C}_P$ (liquid) = 75.5 JK^{-1}mol^{-1} and $\overline{C}_P$ (solid) = 37.7 JK^{-1}mol^{-1}, respectively, for water liquid and solid. Recalculate the table entries in CTQ 14.

5. 10.0 L of an ideal gas at 0°C and 10.0 bar are expanded to a final pressure of 1.00 bar $\overline{C}_V = 3/2$ R. Calculate ΔU, ΔH, q, w, and ΔS if the process is:

    a)   Reversible and isothermal

    b)   Irreversible and adiabatic

6. Using the results of CTQ 16 write a general relation between dS, dq, and T that applies to reversible and spontaneous processes.

7. Consider a process in which $\Delta_r \overline{C}_P > 0$ and is temperature-independent. The pressure is held constant. If the temperature is raised, does the value of $\Delta_r S$ increase, decrease, stay the same, or is it impossible to determine? Explain your reasoning.

ChemActivity T8

The Third Law of Thermodynamics

Focus Question: Why are all entropies positive?

Model 1: Absolute Zero (0 K).

The lowest possible temperature is absolute zero (0 K). Any system at this temperature has the least amount of energy that the system can possibly attain. If the species present are capable of vibrating, there will be some residual vibrational motion. As the temperature is raised above 0 K, increased motion of the particles composing the system becomes possible.

Critical Thinking Questions

1. Explain why all materials must be solids at absolute zero.

2. As the temperature associated with a system is raised above 0 K, do you expect that the entropy of the system will increase or decrease? Explain your reasoning.

Model 2: The Third Law of Thermodynamics.

The entropy of a perfect crystal is zero when the temperature of the crystal is equal to absolute zero (0 K).

The third law of thermodynamics defines zero on the entropy scale.

Figure 1: Entropy vs. T for a Typical Pure Substance

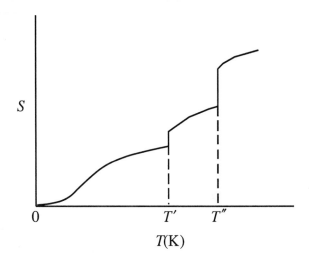

Critical Thinking Questions

3. Explain why the entropy of any system must have a nonnegative value based on the Third Law of Thermodynamics.

4. Note that in Figure 1, the entropy changes gradually from 0 K then undergoes an abrupt change. Explain this result.

5. There are two vertical lines in Figure 1 - at temperatures T' and T". To what physical phenomenon do each of these vertical lines correspond?

6. Label the appropriate portions of Figure 1 corresponding to the solid, liquid, and gas phases.

Model 3: Absolute Entropy.

As temperature is raised above absolute zero, energy is stored by translational, rotational, and vibrational modes and through intermolecular interaction. Thermal entropies are a measure of this increased sharing of energy above that at absolute zero. Heating a substance always increases its entropy because more ways of storing the heat energy become available.

The entropy of diamond is less than that of graphite. The energy levels in diamond are more widely spaced than in graphite and therefore graphite has more ways to store energy and a higher entropy.

The entropy change associated with a change in temperature from T_1 to T_2 within any single phase can be determined from the expression:

$$S_{T_2}^{\circ} - S_{T_1}^{\circ} = \int_{T_1}^{T_2} \overline{C_P} \frac{dT}{T} \tag{1}$$

Because $S_0^{\circ} = 0$ by the Third Law, an absolute entropy for a solid can be defined:

$$S_T^{\circ} - S_0^{\circ} = \int_0^T \overline{C_P} \frac{dT}{T} = S_T^{\circ} \tag{2}$$

This is the excess of entropy over that possessed by the solid at 0 K and arises from thermal energy input.

The entropy change for the phase transition from solid to liquid is the entropy of fusion, $\Delta_r S_{fus}^{\circ}$. The entropy change for the phase transition from liquid to gas is the entropy of vaporization, $\Delta_r S_{vap}^{\circ}$.

Critical Thinking Questions

7. What would you expect the entropy of $H_2O(s)$ to be at 0 K and 1 bar?

8. Provide an expression for the calculation of the entropy of $H_2O(s)$ at 125 K relative to that at 0 K.

9. Consider H_2O at 273.15 K and 1 bar.

    a)   Provide an expression for the calculation of the entropy of $H_2O(s)$ under these
         conditions.

    b)   Is the entropy of $H_2O(\ell)$ under these same conditions greater than, less than,
         or equal to that of $H_2O(s)$? Explain.

    c)   Provide an expression for the calculation of the entropy of $H_2O(\ell)$ under these
         conditions in terms of $\overline{C}_P$, T, and $\Delta H°_{fus}$ for H_2O.

Exercises

1. For the following process show how to calculate the absolute entropy of N_2 at 298 K
    and 1 bar.

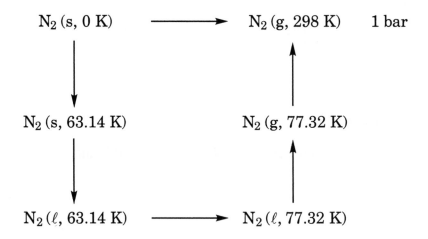

2. A gas confined to a piston-cylinder is expanded by the addition of heat while P is
    maintained constant and a weight is raised. List everything that has its energy altered
    as a result of this process. List everything that has its entropy altered by this process.

3. The entropy of $H_2O(s)$ is not zero at absolute zero temperature. That is, $H_2O(s)$ does
    not obey the Third Law of Thermodynamics. Can you suggest a reason for this?

4. Is it possible for $\Delta H°_{vap} < 0$ for any pure substance? Explain why or why not.

Model 4: The Standard-State Entropy of Reaction.

The **standard-state entropy of reaction, $\Delta_r S°$,** is the change in the entropy of a system that accompanies a chemical reaction measured under **standard-state conditions**. For solids and liquids the standard state is the pure substance at 1 bar. For gases, the standard state is the ideal gas at 1 bar.

Table 1: Absolute Entropies of Various Species

	$S°_{298}$ (JK^{-1}mol^{-1})
$KClO_4(s)$	151.0
$KCl(s)$	82.67
$O_2(g)$	205.03

Critical Thinking Questions

10. What is the meaning of the symbol $S°_{298}$?

11. Why are all of the entries in Table 1 positive?

12. Consider the reaction $KClO_4(s) = KCl(s) + 2 O_2(g)$.

    a) Without referring to Table 1, predict whether $\Delta_r S°$ for the reaction is expected to be positive or negative. Explain your reasoning.

    b) Describe, in words, how $\Delta_r S°$ could be calculated.

    c) Use the data in Table 1 to calculate $\Delta_r S°$.

Exercises

5. For the reaction

$$C(graphite) + O_2(g) \rightleftharpoons CO_2(g)$$

the following data are known:

	$\Delta_f H^{\circ}_{298}$ (kJmol^{-1})	S°_{298} (JK^{-1}mol^{-1})	$\overline{C}_p$(JK^{-1}mol^{-1})
C(graphite)	0	5.74	8.50
$O_2(g)$	0	205.14	29.38
$CO_2(g)$	−393.51	213.74	37.13

a) Calculate $\Delta_r H^{\circ}$ and $\Delta_r S^{\circ}$ at 298 K.

b) Find ΔS_{tot} at 298 K.

c) Find $\Delta_r H^{\circ}$, $\Delta_r S^{\circ}$, and ΔS_{tot} at 400 K.

d) Calculate w for this reaction and $\Delta_r U^{\circ}$ at 298 K.

e) What is $\Delta_r S^{\circ}_0$? State all assumptions.

ChemActivity T9

Gibbs Energy and Helmholtz Energy

Focus Question: Suppose a given chemical mixture has the potential to produce products so that the sum of the bond strengths is larger than those of the reactants but that the number of moles of reactant gases are decreased. Will the reaction occur?

Model 1. Some Thermodynamic Relationships and Definitions.

$$\mathrm{d}S \geq \frac{dq}{T} \tag{1}$$

$$A \equiv U - TS \qquad \text{Helmholtz Energy} \tag{2}$$

$$G \equiv A + PV \qquad \text{Gibbs Energy} \tag{3}$$

Critical Thinking Questions

1. Under what conditions is $dS = \frac{dq}{T}$?

2. Under what conditions is $dS > \frac{dq}{T}$?

3. Use the First Law and equation 1 to obtain an expression showing how work, dw, is related to dS, T, and dU.

4. Consider the Helmholtz energy, defined in equation 2:

    a)   Find dA.

    b)   Find dA for a constant temperature process.

    c)   How is the total work related to ΔA for a constant temperature process?

5. Consider a spontaneous process which takes place at constant temperature and constant volume and no work is done. What is the condition on ΔA for a spontaneous process at constant temperature and volume?

6. Why have thermodynamicists defined a quantity as $U - TS$?

7. Consider the Gibbs energy, G.

    a)   Based on its definition (equation 3), how do we know that G is a state function?

b)   Show that $G = U - TS + PV$ and provide an expression for dG.

c)   Show that $G = H - TS$ and provide an expression for dG.

d)   Find ΔG for a constant temperature process in terms of H and S.

Information

Often, chemical processes are studied under conditions of constant temperature and pressure. The total work for a process can be thought of as the sum of the pressure-volume work and the other (non-pv) work:

$$w = w_{pv} + w_{nonpv} \qquad (4)$$

Critical Thinking Questions

8. For a constant temperature and pressure process, show that

$$\Delta A + P\Delta V \leq w_{nonpv}$$

9. Show how w_{nonpv} is related to ΔG at constant T,P.

10. What is the condition on ΔG for a spontaneous process at constant T,P when $w_{nonpv} = 0$?

11. Why have thermodynamicists defined a quantity as $H - TS$?

Information

$$\Delta S_{tot} = \Delta S_{sys} + \Delta S_{surr} \geq 0 \tag{5}$$

$$\Delta S_{sys} = S_{final} - S_{init} \tag{6}$$

Critical Thinking Questions

12. Use equation 5 to obtain an expression for ΔS_{tot} in terms of ΔS_{sys} and ΔH_{sys} for a process at constant T and P.

13. Give ΔH_{sys} in terms of ΔU_{sys}, pressure, and volume at constant T, P.

14. Given your answers to CTQs 12 and 13:

    a) Rewrite the Second Law ($\Delta S_{tot} \geq 0$) in terms of ΔS_{sys}, ΔU_{sys}, P, V, and T at constant T,P.

b)   Use equation 6 (and analogous expressions for the other variables) to rewrite your answer to CTQ 14a so that all terms referring to final states of the system are on one side of the expression, and all terms referring to initial states are on the other side.

c)   Use the definition of G to rewrite your answer to CTQ 14b in terms of G only.

15.   Given your answer to CTQ 14c:

a)   If $G_{final} > G_{init}$ (at constant T,P), what can be said about the process?

b)   If $G_{final} < G_{init}$, what can be said about the process?

c)   If $G_{final} = G_{init}$, what can be said about the process?

16.   Show that when T and P are constant, $\Delta G = -T\Delta S_{tot}$.

Exercises

1. At T,P constant what does it mean in terms of Gibbs energy to say that:

    a) Ice melts at 0°C and 1 bar?

    b) Water boils at 100°C and 1 bar?

    c) The vapor pressure of water at 20°C is 17.36 torr?

    d) A 1.0 molal aqueous glucose solution freezes at –1.86°C and 1 bar?

    e) The solubility of I_2(s) in water at 50°C is 0.078 gL^{-1} at 1 bar?

    f) $CaCO_3$(s) $= CO_2$(g) $+$ CaO(s)?

2. What does it mean if for a given process at constant T,P, $\Delta G > 0$? $\Delta G < 0$? $\Delta G = 0$?

3. What does it mean if for a given process at constant T,V, $\Delta A > 0$? $\Delta A < 0$? $\Delta A = 0$?

ChemActivity T10

Gibbs Energy as a Function of Temperature and Pressure

Focus Question: **The most stable phase of a substance at a fixed *T,P* is that phase with the lowest molar Gibbs energy. If temperature is increased what happens to the Gibbs energy? What happens to the Gibbs energy if pressure is increased?**

Model 1: The Gibbs Energy.

$$G \equiv U + PV - TS \tag{1}$$

Critical Thinking Questions

1. Describe the meaning of equation 1 using grammatically correct English sentences.

2. Based on equation 1 and the First Law, derive equation 2 below for a reversible process in which the only work done is pressure-volume work:

$$dG = VdP - SdT \ . \tag{2}$$

Clearly state the assumptions and conditions which are necessary for this derivation.

3.   Although the relationship $dG = VdP - SdT$ was derived in CTQ 2 under particular conditions, it is generally applicable, even when those conditions do not apply. Why is this so?

4.   Gibbs energy can be expressed as a function of T and P.

   a)   What is the total differential for $G = G(T,P)$?

   b)   Use equation 2 to find the derivatives $\left(\dfrac{\partial G}{\partial P}\right)_T$ and $\left(\dfrac{\partial G}{\partial T}\right)_P$.

   c)   Express $\left(\dfrac{\partial G}{\partial T}\right)_P$ in terms of G, H, and T.

Exercise

1.   Consider a sample of CaO(s) at 298 K and 1 atm.

   a)   Does the Gibbs energy of the sample increase, decrease, or remain constant if the temperature is raised to 350 K at 1 atm? Explain.

   b)   Does the Gibbs energy of the original sample increase, decrease, or remain constant if the pressure is increased to 2 atm at 298 K? Explain.

Model 2: Chemical Potential.

For a pure phase, the chemical potential is given by

$$\mu = \frac{G}{n} = \overline{G} \tag{3}$$

where $\overline{G}$ is the molar Gibbs energy and n is the number of moles.

Critical Thinking Questions

5. For a constant temperature process, what is the relationship between dG, volume, and pressure?

6. A particularly useful relationship is that between G and volume and pressure for various systems at constant temperature.

    a)   For condensed phases, V can be assumed to be constant over a small pressure interval. For this case, integrate your expression from CTQ 5 from $G°$ to G and from $P°$ to P to obtain a relationship describing how free energy for a condensed phase is related to volume and pressure at constant temperature.

    b)   Combine the ideal gas law with your expression from CTQ 5 and then integrate from $G°$ to G and from $P°$ to P to obtain an analogous relationship for n moles of an ideal gas at constant temperature.

c)   Use your result from CTQ 6b and the definition of chemical potential to write an expression for μ for an ideal gas at a given temperature in terms of $\mu°$, $P°$, P, and T.

7.   Based on your answer to CTQ 6c, how does the chemical potential of an ideal gas change (increase, decrease?) as its pressure is increased above $P°$ (at a given T)?

Model 3: A Mixture of Ideal Gases.

A container is divided by a Pd membrane permeable only to $H_2(g)$. Pure $H_2(g)$ is on one side and a mixture of $H_2(g)$ and $N_2(g)$ is on the other side. All gases are assumed to be ideal.

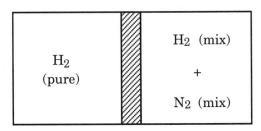

Critical Thinking Questions

8.   When the system in Model 3 is at equilibrium:

a)   What must be the relationship between $\mu_{H_2(pure)}$ and $\mu_{H_2(mix)}$?

b)   What must be the relationship between $P_{H_2(pure)}$ and $P_{H_2(mix)}$?

9.   Provide an expression for $\mu_{H_2(pure)}$ in terms of $\mu_{H_2}^{\circ}$, temperature, and pressure.

10.   Provide an expression for $\mu_{H_2(mix)}$ and then generalize this expression for any gas in an ideal mixture.

Exercise

2.

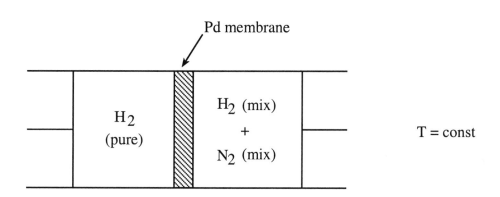

a)   What is the relationship of $\mu_{(pure)}$ to μ° when $P_{H_2(pure)} = 1$ bar?

b)   If $P_{H_2(mix)} = 0.5$ bar and $P_{H_2(pure)} = 1$ bar, on which side of the membrane is the free energy of hydrogen the lowest? In which direction will there be a gas flow?

c)   If $P_{H_2(mix)} = 1$ bar and $P_{H_2(pure)} = 0.5$ bar, on which side of the membrane is the free energy of hydrogen the lowest? In which direction will there be a flow of gas?

ChemActivity T10A
Euler's Criterion

Model: Euler's Criterion.

The thermodynamic state functions U, H, S, A, V, T, P and G are all exact differentials. An exact differential must satisfy Euler's criterion which states that if a function F is exact and $dF = x_1 dy_1 + x_2 dy_2$ then $\left(\dfrac{\partial x_2}{\partial y_1}\right)_{y_2} = \left(\dfrac{\partial x_1}{\partial y_2}\right)_{y_1}$.

Exercises

1. Verify that:

   a) $dG = -SdT + VdP$

   b) $dA = -SdT - PdV$

2. The derivatives $\left(\dfrac{\partial S}{\partial V}\right)_T$ and $\left(\dfrac{\partial S}{\partial P}\right)_T$ can be expressed in terms of derivatives of pressures, volumes, and temperatures. Show how this can be done.

3. From $dU = TdS - PdV$ find $\left(\dfrac{\partial U}{\partial P}\right)_T$.

   Find $\left(\dfrac{\partial U}{\partial V}\right)_T$ for an ideal gas and for a vdW gas. What is $\left(\dfrac{\partial U}{\partial P}\right)_T$ for an ideal gas? Can you generalize what variables affect U for an ideal gas?

4. Show $C_P - C_V = \left[\left(\dfrac{\partial U}{\partial T}\right)_P + P\left(\dfrac{\partial V}{\partial T}\right)_P\right] - \left(\dfrac{\partial U}{\partial T}\right)_V$

   and find the difference between C_P and C_V for an ideal gas. Hint: Find C_P from the definition of H.

ChemActivity T11

Equilibrium

Focus Question: One mole each of $PCl_3(g)$, $Cl_2(g)$, and $PCl_5(g)$ are placed into an evacuated, fixed volume container and maintained at 298 K. What will happen to the pressure in the container as time passes?

$$PCl_5(g) \rightleftharpoons PCl_3(g) + Cl_2(g)$$

Model 1: A Chemical Reaction.

Initially, there are n_{o,NH_3},; n_{o,N_2},; n_{o,H_2}, moles of gas present in some system. The reaction

$$N_2(g) + 3 H_2(g) \rightleftharpoons 2 NH_3(g) \tag{1}$$

proceeds by an amount ξ from left to right.

Critical Thinking Questions

1. If 1.0 mole of reaction 1 occurs:

    a) How many moles of N_2 are consumed?

    b) How many moles of H_2 are consumed?

    c) How many moles of NH_3 are produced?

2. Give the number of moles of each species (n_i) present after ξ moles of reaction.

Information

The rate of change of moles present with respect to ξ, $\dfrac{dn_i}{d\xi}$, is given the symbol v_i.

Critical Thinking Question

3. Consider the rate of change of moles for each of the chemical species in reaction 1.

   a) Evaluate $\dfrac{dn_i}{d\xi} = v_i$ for each of the three species.

   b) How is the sign of v_i related to the nature of component i in reaction 1?

   c) How is the magnitude of v_i related to component i in reaction 1?

   d) Provide a generalization of your responses to CTQs 3b and 3c and provide a brief explanation.

Model 2: Gibbs Energy of Reaction.

For a general reaction

$$aA + bB \rightleftharpoons cC \tag{2}$$

as the reaction proceeds by an amount ξ, the Gibbs energy changes according to

$$\frac{\partial G}{\partial \xi} = c(\Delta_r G_f)_C - a(\Delta_r G_f)_A - b(\Delta_r G_f)_B = \Delta_r G$$

$$= c\mu_C - a\mu_A - b\mu_B \tag{3}$$

where $\Delta_r G_f$ is the Gibbs energy of formation and $(\Delta_r G_f)_i = \mu_i$.

Or, in general, $\dfrac{\partial G}{\partial \xi} = \sum_i v_i \mu_i = \Delta_r G$

The change in free energy of a system that occurs during a reaction can be measured under any set of conditions. If the data are collected under standard-state conditions, the result is the **standard-state free energy of reaction, $\Delta_r G°$**. For gases the standard state superscript circle represents a hypothetical ideal gas state at 1 bar of pressure. For pure solid or liquid phases the standard state is the actual substance at 1 bar pressure.

Critical Thinking Questions

4. Consider reaction 2 as a constant temperature process involving ideal gases.

    a)   Provide an expression for μ_i for each of the components.

    b)   Find $\Delta_r G$ in terms of μ_i and in terms of the partial pressure P_i. Take advantage of the properties of logarithms to collect all of the P_i in one term as a ratio of pressures.

c)   Identify $\Delta_r G°$ in your answer to CTQ 4b, and then rewrite the expression for $\Delta_r G$ in terms of $\Delta_r G°$ and a term involving a ratio of pressures.

d)   If the reaction is at constant T,P and at equilibrium, what is the magnitude of $\Delta_r G$?

Model 3: The Equilibrium Constant.

For reaction (2) at a given T,

$$\left(\frac{P_C{}^c}{P_A{}^a P_B{}^b}\right)_{eq} = K_P \qquad (4)$$

where K_P is the equilibrium constant for reaction 2 at the temperature T. Because μ for pure solids or liquids is relatively independent of pressure, the equilibrium constant expression does not contain a term for pure solids or liquids. However in calculating $\Delta_r G°$, $\Delta_r H°$, or $\Delta_r S°$ for a chemical reaction, these terms must be taken into account.

Critical Thinking Question

5. Based on your answers to CTQ 4 above, derive an expression for the relationship between $\Delta_r G°$ and K_p at constant T and P.

Information

$$\Delta_r G^\circ = \Delta_r H^\circ - T\Delta_r S^\circ \tag{5}$$

Table 1: Relevant Thermodynamic Data for the Reaction
       C(graphite) + H2O(g) ⇌ CO(g) + H2(g)

	$\Delta_r H^\circ_{f\,298}$ (kJmol^{-1})	S°_{298}(JK^{-1}mol^{-1})	$\overline{C}_p^\circ$ (JK^{-1}mol^{-1})
C(graphite)		5.7	8.64
H$_2$O(g)	-242	189	33.58
CO(g)	-111	198	29.14
H$_2$(g)		131	28.84

Critical Thinking Questions

6. What is $\Delta_r H^\circ_{f\,298}$ for C(graphite)? For H$_2$(g)? Explain your reasoning.

7. Calculate $\Delta_r H^\circ_{298}$, $\Delta_r S^\circ_{298}$, and $\Delta_r G^\circ_{298}$ for the reaction described in Table 1.

8. Provide an expression relating $\Delta_r G$ to $\Delta_r G^\circ$ and ratios of partial pressures of the gaseous components for the reaction referred to in Table 1.

9.   Consider a situation in which the four components for the reaction in Table 1 were mixed together in some container, all in their standard states.

   a)   Calculate the value of $\Delta_r G$ under these circumstances.

   b)   Is this system at equilibrium? If so, explain how you know. If not, explain why not and indicate whether the reaction will proceed to the left or to the right.

10.   What would the ratio of partial pressures need to be in order for the reaction to proceed to the right?

11.   Is it possible to determine if a reaction will proceed to the right from examing the appropriate ratio of partial pressures? If so, explain why. If not, indicate what additional information is needed.

12.   Show that

$$\Delta_r G = RT \ln \frac{Q}{K_P} \qquad\qquad (6)$$

   where Q is a non-equilibrium ratio analogous to K_P.

Exercises

1. Provide an expression for K_P for reaction 1. Why does a term for C(graphite) not appear in the K_P expression for the reaction of Table 1?

2. Calculate the equilibrium constant for the reaction using tabulated data.

$$2\,H_2(g) + O_2(g) \rightleftharpoons 2\,H_2O(g) \quad \text{at 298 K, 1 bar}$$

   Write the equilibrium constant for this reaction in terms of the partial pressures.

3. Calculate the equilibrium constant for

$$CaCO_3(\text{calcite}) \rightleftharpoons CaO(s) + CO_2(g) \text{ at 298 K, 1 bar}$$

   from tabulated data.

4. For the general reaction for ideal gas reactions:

$$K_P = \frac{P_C^c}{P_A^a P_B^b}$$

   Show that $P_i = C_i\,RT$ where C_i is the concentration in mol L^{-1}.

   Show that $P_i = X_i\,P_{tot}$ where X_i is the mole fraction and P_{tot} is the total pressure.

   Find the relationship between K_P, K_c, and K_x.

5. For $COCl_2(g) \rightleftharpoons CO(g) + Cl_2(g)$, $K_P = 0.0444$ at 394.8 °C. Assume that initially n moles of $COCl_2$ and no moles of CO and Cl_2 are present. Find the partial pressures of CO and Cl_2 at 1.00 bar total pressure. Find $\Delta_r G°$.

6. At 25 °C and 1.0 bar total pressure N_2O_4 is 19% dissociated. Calculate K_P. Assume ideal behavior.

$$N_2O_4(g) \rightleftharpoons 2NO_2(g)$$

7. For the reaction treated in Table 1 (assume $\overline{C_P}$ is not a function of temperature):

   a) Calculate $\Delta_r H°$ for the reaction at 125 °C.

   b) Calculate $\Delta_r S°$ for the reaction at 125 °C.

   c) Calculate ΔS_{tot} at 125 °C

   d) Does the reaction proceed at 125 °C and 1.00 bar?

e)   What is $\Delta_r U°$ at 125 °C for the reaction? Assume ideal behavior.

f)   Calculate $\Delta_r G°$ at 125 °C. Does the reaction proceed at 125 °C and 1.00 bar? Compare this prediction to that made in d) and comment.

g)   Calculate the equilibrium constant at 125 °C.

8. For the reaction

$$CaCO_3(s) \rightleftharpoons CaO(s) + CO_2(g)$$

$K_p = P_{CO_2}$ but $\Delta_r G° = \Delta_r G°_{fCO_2}(g) + \Delta_r G°_{fCaO}(s) - \Delta_r G°_{fCaCO_3}$.

Why is it that the partial pressure of the gas only appears in K_p but all the standard Gibbs energies of formation are used to calculate $\Delta G°$? Hint: See ChemActivity 10, Exercise 1.

9. The reaction for the formation of nitrosyl chloride

$$2NO(g) + Cl_2(g) \rightleftharpoons 2NOCl(g)$$

was studied at 25°C. The pressures at equilibrium were found to be

$P_{NOCl} = 1.2$ atm
$P_{NO} = 5.0 \times 10^{-2}$ atm
$P_{Cl_2} = 3.0 \times 10^{-1}$ atm

a)   Calculate the value of K_p for this reaction at 25°C.

b)   If 1.0 atm of $NOCl(g)$ was placed into an evacuated container would a reaction occur?

c)   If 1.0 atm each of $NOCl(g)$, $Cl_2(g)$ and $NO(g)$ were placed into an evacuated container would a reaction occur? If so in which direction would the reaction proceed?

d)   If 1.0 atm of $NOCl(g)$, 5.0×10^{-3} atm of NO and 3.0×10^{-1} atm of Cl_2 were placed into an evacuated container would a reaction occur? If so in which direction would the reaction proceed?

Show all calculations and explain each answer.

ChemActivity T12

Temperature Dependence of the Equilibrium Constant

Focus Question: **Why does the extent of a chemical reaction depend on temperature?**

Model 1: The Relationship Between $\Delta_r G°$ and K.

$$\Delta_r G° = -RT \ln K \tag{1}$$

Critical Thinking Questions

1. Use grammatically correct English sentences to describe in words the meaning of equation 1.

2. Rearrange equation 1 and make appropriate substitutions to provide an expression for $\ln K$ as a function of $\Delta_r H°$, $\Delta_r S°$, and T. This expression will be most useful if it is expressed as the sum of two terms, one involving $\Delta_r H°$ and one involving $\Delta_r S°$.

3. Based on your answer to CTQ 2, how does K change as the temperature is increased for a reaction in which (assume $\Delta_r H°$ and $\Delta_r S°$ are not functions of temperature):

   a) $\Delta_r H° > 0$?

   b) $\Delta_r H° < 0$?

   c) $\Delta_r H° = 0$?

Information

$$\frac{d \ln K}{dT} = \frac{\Delta_r H°}{RT^2} \tag{2}$$

Critical Thinking Questions

4. Show how equation 2 can be derived from your answer to CTQ 2, if $\Delta_r H°$ and $\Delta_r S°$ are not functions of temperature.

5. Is equation 2 consistent with your answers to CTQ 3? Explain.

6. Assume $\Delta_r H°$ is not a function of T and integrate equation 2 between definite limits to provide an expression relating K at two different temperatures to those temperatures.

7. Propose a graphical method for obtaining $\Delta_r H°$ of a reaction by measuring K at several different temperatures. Explain your analysis clearly by identifying the slope and intercept of your graph.

8. Equation (2) may be obtained from equation (1) without the assumption that $\Delta_r H°$ and $\Delta_r S°$ are not functions of temperature. To see how this is done start with equation (1) and solve for ln$K$. Then mathematically ask the question how does ln$K$ vary with temperature. Treat the result as a quotient function and evaluate the expression.

Exercises

1. For

$$H_2(g) + I_2(g) \rightleftharpoons 2\,HI(g)$$

   $K_P = 870$ at $298°K$, and $\Delta_r H^\circ_{298} = -10.38$ $kJmol^{-1}$.

   Find K_P at $328°$. Asssume $\Delta_r H^\circ$ is not a function of temperature.

2. A certain nucleotide, N, specifically activates an enzyme according to

$$Enzyme + N \rightleftharpoons Enzyme - N$$

$t,°C$	K
22	4.58×10^3
38	14.5×10^3

   Find $\Delta_r H^\circ$, $\Delta_r G^\circ$, $\Delta_r S^\circ$ for the reaction at 295K.

3. If $\Delta_r C_P \neq f(T)$ show that $\Delta_r H^\circ$ at any temperature T is related to $\Delta_r H^\circ$ at a known temperature, usually 298 K, by

$$\Delta_r H^\circ_T = \Delta_r H^\circ_{298} + \Delta_r \overline{C}_P\,(T - 298)$$

   Find an expression for $\ln K$ as a function of temperature.

4.

	$\Delta_r H^\circ_{f298}$ $(kJmol^{-1})$	S°_{298} $(JK^{-1}mol^{-1})$
$CH_4(g)$	-74.847	186.19
$H_2O(g)$	-241.826	188.72
$CO(g)$	-110.523	197.91
$H_2(g)$	0	130.58

   C_P is given in ChemActivity 4.

   a) Determine K_P at 298K for

$$CO(g) + 3H_2(g) \rightleftharpoons CH_4(g) + H_2O(g)$$

   b) Find K_P at 500K.

5. Show that $\left.\dfrac{\partial(\Delta A/T)}{\partial T}\right)_V = \dfrac{-\Delta U}{T^2}$.

Temperature and Pressure Dependence of Phase Equilibria for Pure Phases

Focus Question: **Does the melting point of ice increase, decrease, or remain constant when the pressure is increased?**

Information

A phase is a region of space in which the intensive properties vary continuously as a function of position. The intensive properties change abruptly across the boundary between phases. For equilibrium between phases, the chemical potential of any species is the same in all phases in which it exists.

Model 1: Two Phases in Equilibrium.

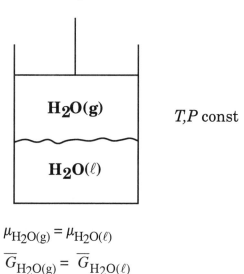

T, P const

$$\mu_{H_2O(g)} = \mu_{H_2O(\ell)}$$

$$\overline{G}_{H_2O(g)} = \overline{G}_{H_2O(\ell)}$$

Critical Thinking Questions

1. Why can the equilibrium condition for the pure phase equilibria of Model 1 be written as

$$\overline{G}_{H_2O(g)} = \overline{G}_{H_2O(\ell)} \ ?$$

2. Show that $d\overline{G} = \overline{V}\,dP - \overline{S}\,dT$ for pure phases where the super bars refer to molar quantities.

Figure 1: Schematic Representation of $\overline{G}$ vs. P for H$_2$O(ℓ) and H$_2$O(g) at 373 K

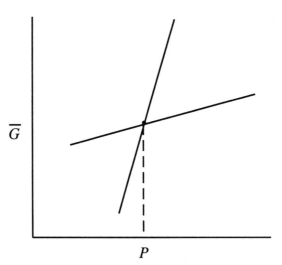

3. In Figure 1:

    a)   Label the appropriate solid line "gas" and the appropriate solid line "liquid". Explain your reasoning.

b)   What is the pressure corresponding to the point where the two lines cross (indicated by the dashed line)?

Exercises

1.   Use Figure 1 to explain why the boiling point of water is decreased at higher elevations.

2.   In Figure 1, the line representing the gas phase should not be straight, expecially over a significant pressure range. Why is this the case? Sketch a graph of $\overline{G}$ vs. P at constant T for a typical ideal gas which shows this curvature, and explain the shape.

Table 1: Standard-State Entropies for H₂O

	$S°$ (J K^{-1} mol^{-1})
$H_2O(s)$	???
$H_2O(l)$	109
$H_2O(g)$	188

$T = 298$ K for liquid and gas. $P = 1$ bar for all phases.

Critical Thinking Questions

4.   For a gaseous phase, $\overline{G}(g) = \overline{H}(g) - T\overline{S}(g)$. Provide an equivalent expression for a liquid phase.

5.   Consider a plot of $\overline{G}$ vs. T for liquid water. Assume that over the temperature interval considered, $\overline{H}$ and $\overline{S}$, are essentially constant.

a)   Based on your response to CTQ 4, what should be the slope of the line representing the graph of $\overline{G}$ vs. T ?

b)   Based on your response to CTQ 4, what should be the y-intercept (value of $\overline{G}$ when $T = 0$) of the line?

c)   On the figure below, <u>sketch</u> a plot for $\overline{G}$ vs. T for liquid water at 1 bar, and then sketch a plot for $\overline{G}$ vs. T for gaseous water at 1 bar on the same figure. Label each of these lines clearly, including the phase and the pressure.

Figure 2: $\overline{G}$ vs. T for H₂O(ℓ) and H₂O(g)

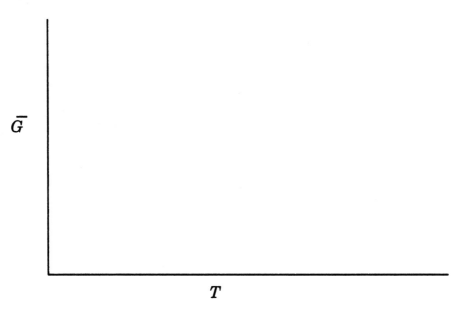

6.   Consider the situation in which H₂O(ℓ) and H₂O(g) are in equilibrium at a particular T and P.

a)   What condition must $\overline{G}(\ell)$ and $\overline{G}(g)$ meet for the two phases to be at equilibrium?

b)   The two lines that you have drawn on Figure 2 should intersect. If they <u>do</u> not, think carefully about your answer to CTQ 6a and the dependence of $\overline{G}$ on T; then redraw the figure.

What is the significance of the temperature at which the two lines cross? Explain.

7. We are now interested in adding additional lines to Figure 2—a plot of $\overline{G}$ vs. T for water at 0.5 bar.

    a) How is $\overline{G}$ for a gas affected by changing the pressure at a given temperature? (Hint: Think about how $d\overline{G}$ and dP are related at a given T.)

    b) Based on your answer to CTQ 7a, in which phase is $\overline{G}$ most dramatically affected by changing the pressure?   Explain.

    c) Sketch a plot of $\overline{G}$ vs. T for gaseous water at 0.5 bar on Figure 2 and label it clearly.

    d) The graph of $\overline{G}$ vs. T for liquid water at 0.5 bar is essentially identical to the graph for liquid water at 1 bar. Label the line in Figure 2 appropriately and explain this result.

    e) What is the significance of the point at which the two lines for $P = 0.5$ bar intersect?

Information

$$d\overline{G} = \overline{V}dP - \overline{S}\,dT \qquad\qquad (1)$$

$$H_2O(\ell) \rightleftharpoons H_2O(g) \qquad\qquad (2)$$

Critical Thinking Questions

8. Consider the situation in which reaction 2 is at equilibrium at some fixed particular T and P.

   a)   What is the relationship between $d\overline{G}(\ell)$ and $d\overline{G}(g)$ at equilibrium?

   b)   What relationship must exist for the temperatures and pressures of these two phases at equilibrium?

9. Consider reaction 2 at equilibrium.

   a)   Use equation 1 to provide an expression relating the volumes, pressures, entropies, and temperatures of the two phases. Use symbols such as $\overline{V}(\ell)$, $\overline{V}(g)$, $\overline{S}(\ell)$, $\overline{S}(g)$, etc. to represent the molar quantities for each phase.

   b)   Rearrange your answer to CTQ 9a to obtain

$$\frac{dP}{dT} = \frac{\Delta_r\overline{S}_{vap}}{\Delta_r\overline{V}} \qquad\qquad (3)$$

c)   Consider the relationship between $\Delta_r\overline{S}$ and $\Delta_r\overline{H}$ for two phases at equilibrium to show that

$$\frac{dP}{dT} = \frac{\Delta_r\overline{H}}{T\Delta_r\overline{V}} \tag{4}$$

d)   For the equilibrium $H_2O(\ell) \rightleftharpoons H_2O(g)$ show that

$$\frac{d \ln P}{dT} = \frac{\Delta_r\overline{H}_{vap}}{RT^2} \tag{5}$$

where $\Delta_r\overline{H}_{vap}$ is the molar enthalpy difference between gas and liquid water and P is the vapor pressure. State all assumptions.

Table 2: Thermodynamic Parameters for the Two Common Forms of CaCO₃, Calcite and Aragonite, at 298 K and 1 atm

	$\Delta_f G_f°$(kJ/mol^{-1})	$\Delta_r H_f°$(kJ/mol^{-1})	$S°$ (JK^{-1}mol^{-1})	V (mL mol^{-1})
Calcite	-1128.8	-1206.9	92.9	36.90
Aragonite	-1127.8	-1207.1	88.7	33.93

Critical Thinking Questions

10. Consider the two common forms of $CaCO_3$ at 298 K and 1 atm.

   a) Which of the two forms of $CaCO_3$ has stronger forces holding the crystal together at this T and P? Explain your reasoning.

   b) Which of the two forms has the most widely spaced and less available energy levels? Explain.

   c) Which of the two common forms of $CaCO_3$ is more stable at this T and P? Explain your reasoning.

Exercises

3. Create a figure analogous to Figure 1 which includes plots for liquid and solid water at 273 K. Label each line carefully, and explain the significance of the temperature at which the two lines intersect.

4. If a wire with weights attached to its ends is placed over a block of ice at its freezing point, the wire will "eat" through the ice. Explain this phenomenon.

5. Integrate the relationship given in CTQ 9d and show how to determine the enthalpy of vaporization of a liquid experimentally.

6. Consider the equilibrium between a pure solid and its vapor and show

$$\frac{d \ln P}{dT} = \frac{\Delta_r \overline{H}°_{sub}}{RT^2}$$

where $\Delta_r \overline{H}_{sub}$ is the enthalpy of sublimation.

7. How could the enthalpy of sublimation of a solid be determined experimentally?

8. For the equilibrium between a solid and liquid show that

$$\frac{dP}{dT} = \frac{\Delta_r \overline{H^\circ}_{fus}}{T \Delta_r \overline{V}}$$

9. Identify all symbols in Exercise 8. Also show that

$$(P_2 - P_1) = \frac{\Delta_r \overline{H^\circ}_{fus}}{\Delta_r \overline{V}} \ln \frac{T_2}{T_1}$$

10. The enthalpy of vaporization of water is 2255 Jg^{-1}. Assume $\Delta_r \overline{H}_{vap}$ to be independent of temperature and pressure and calculate the vapor pressure of water at 80°C.

11. The normal boiling point of benzene is 80.2°C. The vapor pressure at 25°C is 0.13 atm. Find the enthalpy of vaporization of benzene.

12. The vapor pressure of n-pentane is given by $\ln P_{bar} = 8.630 - 2819.7 \, T^{-1} + 1.855 \times 10^{-3} \, T$. Derive an expression for $\Delta_r \overline{H^\circ}_{vap}$ as a function of temperature and find $\Delta_r \overline{H^\circ}_{vap}$ of n-pentane at its normal boiling point 36.1°C.

13. $\Delta_r \overline{H^\circ}_{fus}$ of water is 6024 Jmol^{-1} at 273 K. The densities of solid and liquid water are 0.915 and 1.000 g mL^{-1}, respectively. What is the freezing point of water under 100 bar pressure?

14. $\Delta_r \overline{H^\circ}_{vap}$ and $\Delta_r \overline{H^\circ}_{fus}$ of water are 44.8 KJmol^{-1} and 6024 Jmol^{-1}, respectively at 0°C. The vapor pressure of $H_2O(\ell)$ at 0°C is 4.58 torr. What is the vapor pressure of $H_2O(s)$ at −15°C? State all assumptions.

15. Roughly sketch a plot of $\overline{G}$ vs. T for the two forms of $CaCO_3$. Sketch $\overline{G}$ vs. T for the two forms under very high pressure. At what temperature at 1 bar are aragonite and calcite in equilibrium?

16. At 298 K what pressure is necessary to bring calcite and argonite into equilibrium?

ChemActivity **T14**

Vapor Pressure and One Component Phase Diagrams

Focus Question: Sufficient water is placed in an evacuated container to insure that the liquid phase is always present. At what temperature will the water boil?

Model 1: Boiling of a Liquid.

We recognize that a liquid is boiling by the formation of bubbles in the liquid. These bubbles are balls of vapor formed by molecules that have acquired sufficient energy to enter the vapor phase. If the external pressure, usually the atmospheric pressure, is greater than the vapor pressure developed in these gaseous pockets, the pockets will be crushed and no visible evidence of boiling will be seen. If, however, the vapor pressure in the bubbles is equal to that of the external pressure they will not collapse and will be seen to rise to the surface and then enter the vapor phase above the liquid. Thus, a liquid boils when the pressure of the gas escaping from the liquid is equal to the pressure exerted on the liquid by its surroundings.

Figure 1: Vapor Pressure of H₂O vs. Temperature

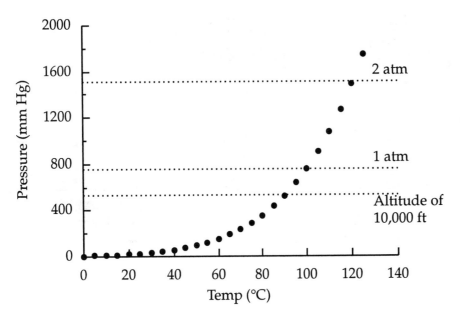

J.N. Spencer, G.M. Bodner, and L.R. Rickard, *Chemistry: Structure and Dynamics*, John Wiley & Sons, Inc., 2nd Ed., 2002, Chpt. 7. This material is used by permission of John Wiley & Sons, Inc.

Critical Thinking Questions

1. When H_2O boils at 1 atm external pressure, what is the vapor pressure of H_2O?

2. What is the temperature at which the vapor pressure of H_2O is 2 atm?

3. What is the boiling point of H_2O at 2 atm?

4. If you like a really hot cup of tea would you prefer to live up in the Rocky Mountains above Denver or in Philadelphia? Explain your reasoning.

Figure 2: Water in a Piston-Cylinder at Various Temperatures

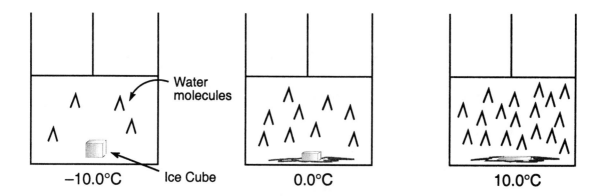

−10.0°C   Ice Cube   0.0°C   10.0°C

Table 1: Vapor Pressure of H₂O (under the pressure of its own vapor)

	$t,°C$	VP (Torr = mmHg)
$H_2O(s)$	−10.0	1.95
	0.0	4.58
$H_2O(\ell)$	0.0	4.58
	10.0	9.2
	50.0	92.5
	80.0	355.1
	95.0	633.9
	100.0	760.0

Critical Thinking Questions

5. Sufficient ice is placed into an evacuated, fixed volume, piston-cylinder container to ensure that ice is always present. If the temperature is held at -10.0 °C, describe the final state of the system by indicating

    a) how many phases will coexist.

    b) what the pressure will be in the container.

        Explain your reasoning.

6. Suppose that the volume of the system described in CTQ 5 is maintained, but that the temperature is raised to 0.0 °C.

  a)   How many phases are present?

  b)   What is the pressure in the container?

  c)   Assuming that the representation of gas molecules in Figure 2 at -10.0 °C is proportional to the actual number of molecules present, explain why the representation at 0.0 °C is appropriate.

7. The system is maintained at the volume described in CTQ 5, and the temperature is increased to 10.0 °C.

  a)   How many phases are present?

  b)   What is the pressure in the container?

  c)   Explain the representation in Figure 2 for 10.0 °C.

  d)   How many $\wedge$ would be needed to represent the situation at 100 °C at this same volume?

8. Suppose that the system is returned to the volume and temperature (-10.0 °C) of CTQ 5. The piston is then pulled out to increase the volume of the container. The temperature remains constant.

    a)   Will the number of molecules in the gas phase increase, decrease, or remain the same as the piston is pulled out? Explain.

    b)   If the piston is pulled out to double the volume of the container, what will be the final pressure of the system? Explain.

    c)   How does the final pressure depend on the volume of the container?

9. In which, if any, of the scenarios described in CTQs 5–8 does the liquid water boil? Explain your reasoning using grammatically correct English sentences.

Model 2: Behavior of Water at a Total Pressure of 1 Atm.

Piston-Cylinders at -10.0 °C and 1 Atm Total Pressure

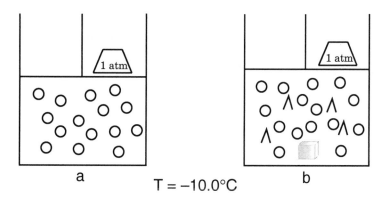

a   T = −10.0°C   b

Dry air, represented by the circles in Model 2, is added to a piston-cylinder container to produce a pressure of 1 atm (1.01 bar) at -10.0 °C. A cube of ice is then added to the container, and a total pressure of 1 atm is maintained by adjustment of the piston.

Each circle in Model 2 represents about 50 Torr of air molecules in the container. The water molecules are represented as in Figure 2.

Critical Thinking Questions

10.   Why is the piston on the right slightly higher than the piston on the left?

11.   Draw piston-cylinder representations of the system as in Model 2b for temperatures of 0.0 °C and 10.0 °C.

12.   When the system described in Model 2b undergoes a temperature increase from -10.0 °C to 0.0 °C, does the partial pressure due to water molecules in the container increase, decrease, or remain constant? Explain your reasoning.

13.   Describe the system at 0.0 °C and 1 atm total pressure by indicating the phases present and the partial pressures of any vapor species.

14.   Describe the system (as in CTQ 13) at 10.0 °C and 1 atm total pressure.

15. At 100 °C and 1 atm pressure, the piston-cylinder diagram representation would contain too many water molecules to draw. Explain why this is the case. Your explanation should include a description of the state of the system under these conditions.

16. In which, if any, of the scenarios described in Model 2 does the liquid water boil? Explain your reasoning using grammatically correct English sentences.

Figure 3: Schematic One Component Phase Diagram for H₂O (not drawn to scale)

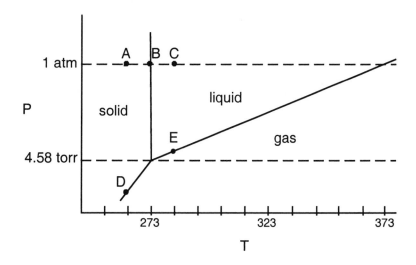

Critical Thinking Questions

17. For each of the lettered points in Figure 3,

    a)   find the system described previously in this ChemActivity which is best represented by that point.

b)   describe the state of the system at each point.

18. What is the significance of

a)   the line separating the region labeled "liquid" from the region labeled "gas"?

b)   the line separating the region labeled "solid" from the region labeled "liquid"?

c)   the line separating the region labeled "solid" from the region labeled "gas"?

19. What determines the slope of the solid lines in the phase diagram?

20. In the region labeled "solid", is the only phase present the solid phase? Explain.

21. In the region labeled "liquid", is the only phase present the liquid phase? Explain.

22. In the region labeled "gas", is the only phase present the gaseous phase?

23. The point at which the three solid lines intersect is known as a "triple point". Describe the state of the system at the triple point.

Exercises

1. A flask partially full of liquid water is heated to boiling and promptly capped. Cold water is poured over the flask. The water boils even more vigorously. Explain.

2. Given the following data roughly sketch the phase diagram for CO_2.

   At –78.5°C the vapor pressure of solid carbon dioxide is 760 Torr. The triple point is at 216.55 K and 5.112 atm. The melting point at 5.2 atm is –56.6°C. The enthalpy of fusion is 180.7 Jg^{-1} and the enthalpy of vaporization is 526.6 Jg^{-1}. The density of the solid is 1.51 gcm^{-3} and that of the liquid is 1.18 gcm^{-3}.

3. The vertical line in Figure 3 is not absolutely vertical. It actually has a negative slope - a very large negative slope. Based on this information, what happens to the freezing point of water as the pressure is increased?

4. What is the effect of an external applied pressure on vapor pressure? Is the vapor pressure of ice in Figure 2 at –10°C the same as that in Figure 3 at the same temperature? Hint: Proceed as in CA T13 CTQ 8,9 but recognize that T is constant.

5.

Starting at 1 bar on the above diagram describe the state of the system at each lettered point.

At a lower constant pressure describe the state of the system at points G, H, and I.

The Ideal Solution

Focus Question: **An equi-molar mixture of benzene and toluene is prepared. What will be the composition of the vapor in equilibrium with this solution?**

Model 1: Benzene and Toluene in the Vapor Phase.

Figure 1: A Mixture of Benzene and Toluene in the Vapor Phase Behaves as a Mixture of Ideal Gases

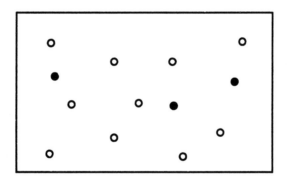

o benzene

● toluene

Recall that for a mixture of ideal gases,

$$P_i V = n_i RT \qquad \text{for each component } i \tag{1}$$

$$P_{tot} V = n_{tot} RT \tag{2}$$

The partial pressure, P_i, of each component in a mixture of gases is related to the composition of the vapor phase according to the relation

$$P_i = X_{i(\text{vap})} P_{tot} \tag{3}$$

where $X_{i(vap)}$ is the mole fraction of component i in the vapor phase. Equation (3) is known as Dalton's Law.

Critical Thinking Questions

1. Show how equation (3) can be derived from equations (1) and (2).

2. At a given temperature and volume, does the partial pressure of benzene, P_{bz}, in Figure 1 depend on the *number* of moles of benzene present in the gas phase? Explain.

Figure 2: Benzene and Toluene in Equilibrium with the Vapor Phase at 300 K

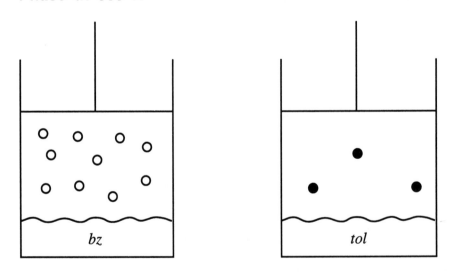

Liquid benzene (*bz*) and liquid toluene (*tol*) at 300 K each in equilibrium with its gas phase.

Information

The vapor pressure of a liquid may be thought of as a measure of the tendency of the molecules to escape into the gas phase. This tendency is directly related to the strength of the interactions in the liquid phase.

Critical Thinking Questions

3. Draw Lewis structures for benzene (C_6H_6) and toluene ($C_6H_5CH_3$).

4. Provide a list of the 4 principal types of intermolecular interaction present between molecules.

5. What type of intermolecular interaction is expected to be the dominant interaction between:

    a)   two benzene molecules (*bz-bz*)?

    b)   two toluene molecules (*tol-tol*)?

    c)   a benzene molecule and a toluene molecule (*bz-tol*)?

6. Based on Figure 2, which species (benzene or toluene) has a higher vapor pressure at 300 K? Explain your reasoning.

7. Based on Figure 2, which is the stronger intermolecular interaction: *bz-bz* or *tol-tol*? Explain.

Table 1: Partial Vapor Pressures of Benzene and Toluene for Various Mixtures at 300 K

moles of bz	moles of tol	$X_{bz(sol)}$	P_{bz} (Torr)	P_{tol} (Torr)	P_{tot} (Torr)
1.00	0	1	103.01	0	
0.00	1.00	0	0	32.1	
0.200	1.80		10.3	28.9	
0.400	1.60		20.6	25.7	
0.800	1.20		41.2	19.2	
0.100	0.900		10.3	28.9	
0.800	0.200		82.4	6.4	

$X_{bz(sol)}$ is the mole fraction of benzene present in the liquid solution.

Critical Thinking Questions

8. Complete Table 1 by calculating the missing values for $X_{bz(sol)}$ and P_{tot} .

9. What is the vapor pressure of pure benzene, P_{bz}^*?

10. What is the vapor pressure of pure toluene, P_{tol}^*?

11. Are your answers to CTQs 9 and 10 consistent with Figure 2? Explain your reasoning.

12. Is P_{bz} determined by the *number of moles* of benzene present in liquid solution? Explain your reasoning.

13. Recall that Dalton's Law describes the relationship between the partial pressure of a component and the composition of the *vapor phase*.

     Use Table 1 to find the relationship between the partial pressure of benzene over the solution and the composition of the *solution*. Provide an answer to this question as both a grammatically correct English sentence *and* as a mathematical relationship.

14. What is the relationship between the partial pressure of toluene over the solution and the compostition of the *solution*? Provide an answer to this question as both a grammatically correct English sentence *and* as a mathematical relationship.

15. Construct a diagram similar to those in Figure 2 (having the same gas phase volume) representing a mixture of 0.8 moles of liquid benzene and 1.2 moles of liquid toluene in equilibrium with its vapor at 300 K.

     Is the composition of the vapor phase the same as the composition of the liquid phase?

Model 2: The Ideal Solution.

A liquid mixture of (at least) two substances is referred to as an ideal solution when Raoult's Law is obeyed by every component. In this case, the volumes are also additive (that is, the volume of the mixture is equal to the sum of the volumes of the components).

Raoult's Law states that the partial pressure of each component of a mixture is equal to the mole fraction of the component in solution multiplied by the vapor pressure of the component when pure.

When substances are mixed the volumes tend to be additive if:

a)   the species mixed are roughly the same size and

b)   the dominant type of intermolecular interaction between the two different species is similar to the dominant type of interaction between the molecules of each pure component.

Critical Thinking Questions

16.   Provide a mathematical representation of Raoult's Law.

17.   Is the mixture of benzene and toluene likely to be an ideal solution? Explain.

18.   Propose a liquid (other than toluene) to mix with benzene which you would *not* expect to result in an ideal solution and explain your reasoning.

Exercises

1.   Would you expect a mixture of dibutyl ether and H_2O to be an ideal solution? Why or why not?

2.   Although Raoult's Law and Dalton's Law are very similar in form, they are in fact very different. Write both laws and carefully describe the differences between them.

3.   Assuming that they behave as ideal gases, calculate the $X_{bz(vap)}$ and $X_{tol(vap)}$ for the mixtures given in Table 1.

4. At 20°C the vapor pressures of pure toluene and xylene are 25 and 5 Torr, respectively. What is the composition of the vapor phase in equilibrium with a solution containing 1.0 mol of toluene and 1.0 mol xylene?

5. The following relationship is particularly useful for relating the composition of an ideal solution to the vapor phase composition.

$$X_{A(vap)} = \frac{X_{A(sol)}P_A^*}{X_{A(sol)}P_A^* + (1-X_{A(sol)})P_B^*}$$

Derive this relation from Raoult's and Dalton's laws.

6. Is it possible for the composition of the vapor phase and the composition of the liquid phase to be identical for a mixture of two liquids? If so, under what conditions? If not, why not?

ChemActivity T16

Chemical Potential for a Component of a Solution

Focus Question: Addition of a solute to a solvent increases the solvent's contribution to the entropy of the liquid phase. Does the solvent's contribution to the Gibbs energy of the liquid phase increase or decrease?

Model 1: Gibbs Energy of a Pure Liquid and its Vapor.

Consider pure liquid A in equilibrium with its vapor. When phases are in equilibrium, if there is a gas phase it is called the vapor(*vap*) phase.

Figure 1: Pure Liquid A at 298 K in Equilibrium with its Vapor

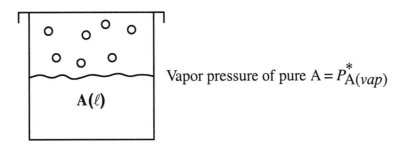

Vapor pressure of pure A $= P^*_{A(vap)}$

$A(\ell)$

When two or more phases are in equilibrium at constant temperature and pressure, the temperatures and pressures of all phases must be the same and the Gibbs energy of each component present in each phase must be identical.

Information

The Gibbs energy of pure liquid A is $\mu_A^*(\ell)$.
The Gibbs energy of pure gas A at 1 bar is $\mu_{A(g)}^{\circ}$.

μ_i is the partial molar Gibbs energy and is frequently called the *chemical potential*. It is a measure at constant T and P of the escaping tendency of a component from a phase. Various symbols are used to represent the chemical potential, μ_i and $\overline{G}_i$ being the most common.

Critical Thinking Questions

1. For the system at equilibrium described in Figure 1:

    a)   what are the two phases present and what is the composition of each phase?

    b)   what pressure is exerted on A(ℓ)?

    c)   what is the temperature of A(g)?

    d)   what is the mathematical relationship between $\mu_A^*(\ell)$ and $\mu_{A(vap)}$ at equilibrium?

2. Provide an expression for the $\mu_{A(vap)}$ in terms of $\mu_{A(g)}^{\circ}$, the temperature T and the vapor pressure $P_{A(vap)}^*$.

Model 2: An Ideal Mixture of Benzene and Toluene.

Consider an ideal mixture of benzene and toluene at equilibrium with its vapor. Assume that the vapor phase behaves ideally also.

Figure 2: A Mixture of Benzene and Toluene in Equilibrium with the Vapor Phase at 300 K

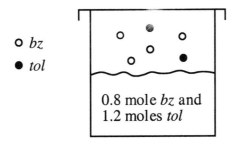

○ *bz*

● *tol*

0.8 mole *bz* and
1.2 moles *tol*

Critical Thinking Questions

3. In Figure 2, what relationship must exist between:

   a) $\mu_{bz(sol)}$ and $\mu_{bz(vap)}$?

   b) $\mu_{tol(sol)}$ and $\mu_{tol(vap)}$?

4.   For the solution in Figure 2 provide an expression similar to that in CTQ 2 for $\mu_{bz(vap)}$, the Gibbs energy of benzene in the vapor. Clearly identify all symbols.

5.   Use Raoult's Law to show that

$$\mu_{bz(sol)} = [\,\mu_{bz(g)}^{\circ} + RT \ln P_{bz(vap)}^{*}\,] + RT \ln X_{bz(sol)} \,. \tag{1}$$

6. Identify each term in the bracketed expression in equation (1), and then explain the significance of the entire bracketed term.

7. Show that equation (1) can be simplified to obtain

$$\mu_{bz(sol)} = \mu_{bz(\ell)}^* + RT \ln X_{bz(sol)} . \tag{2}$$

Clearly define all symbols.

8. Generalize equation (2) to provide an expression for μ_i, the Gibbs energy of any component i in an ideal solution in terms of its mole fraction, X_i.

Exercises

1. $\mu_{i(sol)} = \mu_{i(\ell)}^* + RT \ln X_{i(sol)}$

   Describe in words the meaning of this equation.

2. A and B mix to form an ideal solution. What is the Gibbs energy of A in the solution as compared to the Gibbs energy of pure liquid A? How does the Gibbs energy of B(sol) compare to B(pure)?

3. If B is added to pure A what happens to the Gibbs energy of A?

4. Why does the Gibbs energy of the solvent of an ideal solution decrease upon addition of a solute?

5. Assume that the result from CTQ (8) applies to the (clearly non-ideal) solution of salt in water. Use this result to explain (in grammatically correct English sentences) why the boiling point of H_2O is raised when a small amount of salt is dissolved in it.

Partial Molar Quantities

Focus Question: **When 70.0 mL of water is mixed with 30.0 mL of ethanol, what is the total volume?**

Model 1: Some Properties of Ethanol and Water.

	MW (gmol^{-1})	Density $(25\,^\circ\text{C}, \text{gmL}^{-1})$	$\overline{V}^*$ $(25\,^\circ\text{C}, \text{mLmol}^{-1})$
Ethanol, C_2H_5OH	46	0.78	59
Water, H_2O	18	1.0	18

$\overline{V}^*$ is the molar volume of the pure liquid.

Critical Thinking Questions

1. Show how the molar volume, $\overline{V}^*$, of ethanol can be calculated from its molecular weight and density.

2. By how much does the volume of liquid increase when:

   a)   1.0 mole of water is added to a beaker already containing pure water?

   b)   1.0 mole of ethanol is added to a beaker already containing pure ethanol?

3. Would you expect a mixture of ethanol and water to behave ideally? Why or why not?

4. What would you expect the resulting volume to be if you mixed 30.0 L of ethanol with 70.0 L of water? State any assumptions that you make.

Model 2: Mixtures of Ethanol and Water.

When measured accurately, the volume of a mixture of 70.0 L of water and 30.0 L of ethanol is 96.8 L at 1 atm and 25 °C.

When 1.0 mole of water is added to the above mixture, the volume is observed to increase by 18 mL. When 1 mole of ethanol is added to the mixture, the volume is observed to increase by 52.6 mL. In both cases, the temperature and pressure remain constant.

Critical Thinking Questions

5. Compare the results of adding water and ethanol to the mixture as described in Model 2 to your answers to CTQ 2. Describe similarities and differences.

6. Is the information provided in Model 2 consistent with the mixture behaving ideally? If so, explain how. If not, provide a possible explanation on the molecular level for the observed deviation from ideality.

Model 3: Partial Molar Quantities.

The *partial molar volume* $\overline{V}_{H_2O}$ of water in the mixture above is the change in the total volume of the mixture per mole of added water, at constant T, P and moles of ethanol. This can be described mathematically for a general component i by

$$\overline{V}_i = \frac{\partial V}{\partial n_i}\bigg)_{T,P,n_j} \tag{1}$$

where n_j is the number of moles of component j in the mixture.

Volume is an example of an *extensive* variable - a property or variable that depends on the total amount of substance being considered. Other examples are mass, energy, heat capacity, and entropy. For any extensive property F that can be written as a function of T, P, n_i, n_j, ... the partial molar quantity $\overline{F_i}$ is defined as

$$\overline{F_i} = \frac{\partial F}{\partial n_i}\bigg)_{T,P,n_j} . \tag{2}$$

Further, the property of a mixture, F_{mix}, can be expressed in terms of its partial molar properties as

$$F_{mix} = n_i\overline{F_i} + n_j\overline{F_j} + \; ... \tag{3}$$

Critical Thinking Questions

7. For the mixture described in Model 1,

   a) what is $\dfrac{\partial V}{\partial n_{H_2O}}\bigg)_{T,P,n_{EtOH}}$?

   b) what is $\dfrac{\partial V}{\partial n_{EtOH}}\bigg)_{T,P,n_{H_2O}}$?

8. How is the partial molar Gibbs energy, $\overline{G_i}$ related to the quantity μ_i?

9. For a mixture of two components A and B, provide an expression for G_{mix} in terms of μ_A and μ_B at constant T,P.

10. Using grammatically correct English sentences, describe the meaning of the following relations:

$$\frac{\partial G}{\partial n_A}\bigg)_{T,P,n_B} = \overline{G_A} = \mu_A = \overline{H_A} - T\overline{S_A}$$

Exercises

1. When the contents of a 1000 cubic-foot container filled with bowling balls is mixed with the contents of a 200 cubic-foot container filled with tennis balls, the mixture does not completely fill a 1200 cubic-foot container.

    Explain this result in terms of the partial molar volumes of bowling balls and tennis balls in this mixture. State clearly the assumptions that you make.

2. Is it possible to mix two species without producing a final volume greater than the volume of one species only? Explain.

Information: Maxwell's Relations

In general if dF is an exact differential

$$dF = x_1 dy_1 + x_2 dy_2 + x_3 dy_3 + \cdots$$

then

$$\left(\frac{\partial x_i}{\partial y_j}\right)_{y_k} = \left(\frac{\partial x_j}{\partial y_i}\right)_{y_k}$$

where y_k indicates that all y other than the one considered is constant.

Critical Thinking Questions

11.   Write the total differential for $G = G(T,P,n_A,n_B)$.

12.   Substitute appropriate symbols for all four derivatives.

13.   Use dG in CTQ 11 and Maxwell's Relations to show that

$$\left(\frac{\partial V}{\partial T}\right)_{P,n_A,n_B} = -\left(\frac{\partial S}{\partial P}\right)_{T,n_A,n_B}$$

14. Use Maxwell's Relations to find equivalent derivatives for:

$$\left(\frac{\partial \mu_A}{\partial P}\right)_{T,n_A,n_B} , \quad \left(\frac{\partial \mu_A}{\partial T}\right)_{P,n_A,n_B} , \quad \left(\frac{\partial \mu_B}{\partial T}\right)_{P,n_A,n_B}$$

15. Give the partial molar thermodynamic parameter equivalent of each of the following derivatives:

$$\left(\frac{\partial \mu_i}{\partial T}\right)_{P,n_A,n_B} , \quad \left(\frac{\partial \mu_i}{\partial P}\right)_{T,n_A,n_B} , \quad \left(\frac{\partial (\mu_i/T)}{\partial T}\right)_{P,n_A,n_B}$$

Exercises

3. For an ideal liquid mixture of A and B give the expressions for $\mu_{A(sol)}$ and $\mu_{B(sol)}$. From these expressions find $\overline{H}_A$, $\overline{H}_B$, $\overline{V}_A$, and $\overline{V}_B$. What interpretation can you make concerning what happens to enthalpy and volume when A and B are mixed to form an ideal solution?

   Use your answer to CTQ 15 to find $\overline{S}_A$ and $\overline{S}_B$ for an ideal mixture.

4. For entropy, enthalpy, or Gibbs energy the change produced on mixing can be calculated from $\Delta F_{mix} = F_{mix} - n_A F_{A(pure)} - n_B F_{B(pure)}$ where $F = S$, H, or G and n refers to the numbers of moles mixed. Derive a general expression for ΔG_{mix} and ΔS_{mix} in terms of X_A and X_B.

5. What is ΔH_{mix} and ΔV_{mix} for an ideal solution?

ChemActivity T18

Colligative Properties

Focus Question:

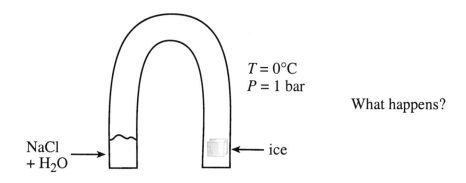

$T = 0°C$
$P = 1$ bar

What happens?

NaCl + H_2O →

ice ←

Model 1: Systems at Equilibrium.

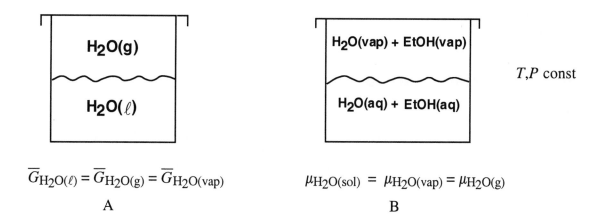

$\overline{G}_{H_2O(\ell)} = \overline{G}_{H_2O(g)} = \overline{G}_{H_2O(vap)}$

A

$\mu_{H_2O(sol)} = \mu_{H_2O(vap)} = \mu_{H_2O(g)}$

B

T,P const

Critical Thinking Questions

1. Why is the symbol $\overline{G}_{H_2O(\ell)}$ used to describe the Gibbs energy per mole in figure A but $\mu_{H_2O(sol)}$ in figure B?

2. What is the relation for Gibbs' energy in Model B for ethanol in the two phases?

3. For model A what variables affect $\overline{G}_{H_2O(g)}$?

4. For model B what variables affect $\mu_{H_2O(vap)}$?

Model 2: One Pure and One Mixed Phase.

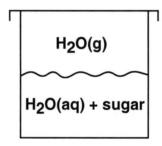

Sugar is a non-volatile solute.

Critical Thinking Questions

5. Plot μ vs. T for water liquid and water gas at constant P on the same graph. Put on this graph a plot of μ vs. T for a solution of sugar in water. Identify the boiling point of pure water and that of the solution. Which is higher?

6. For the boiling solution write the chemical potential expression for the vapor phase at 1 bar. Assume ideal behavior.

7. Write the chemical potential expression for water in the solution for Model 2. Assume ideal behavior.

8. Use the condition for chemical equilibrium and show that

$$\frac{d \ln X_{H_2O(sol)}}{dT} = \frac{-\Delta_r \overline{H}_{vap(H_2O)}}{RT^2}$$

9. a) Integrate the equation in CTQ 8 and show

$$\ln X_{H_2O(sol)} = \frac{\Delta_r \overline{H}_{vap}}{R} \left(\frac{1}{T} - \frac{1}{T_o} \right)$$

where T_o is the normal boiling point of pure water. Assume $\Delta_r \overline{H}_{vap}$ is not a function of T.

b)   Let $\Delta T_{bp} = T - T_0$. Show that when ΔT_{bp} is small, then

$$-\ln X_{H_2O(sol)} \approx \frac{\Delta_r \overline{H}_{vap}}{R} \cdot \frac{\Delta T_{bp}}{T_0^2}$$

10. If sucrose is the sugar, substitute $1 - X_{suc(sol)} = X_{H_2O(sol)}$ and expand in a Taylor series to obtain

$$X_{suc(sol)} \approx \frac{\Delta_r \overline{H}_{vap}}{R} \frac{\Delta T_{bp}}{T_0^2}$$

assuming that X_{suc} is small.

11. Approximate $X_{suc(sol)}$ for a dilute solution and obtain

$$\Delta T_{bp} = \frac{RT_0^2 MW_{H_2O}}{\Delta_r H_{vap}} \cdot m_{suc}$$
$$= K_{bp}\, m_{suc}$$

where $m_{suc} = $ molality sucrose $= \dfrac{\text{moles sucrose}}{1000 \text{ g } H_2O}$

and $MW_{H_2O} = $ molecular weight in kg

and K_{bp} is the boiling point elevation constant for water solvent.

12. Generalize CTQ 11; i.e., in general what is the relationship between ΔT_{bp} and the molality of solute?

13. What conditions must be met in order for the relation derived in CTQ 12 to be valid?

Exercise

1. The boiling point elevation of 50.00 g of CCl_4 containing 1.24 g of an unknown solute is 1.30 degrees. Find the molecular weight of the unknown. $K_{bp} = 5.03$ $°$/molal. $T_{bp} = 76.75°C$.

Model 3: Freezing Point of a Solution.

A solution contains a solute B dissolved in a solvent A. Upon cooling the solution pure A crystallizes out of the solution.

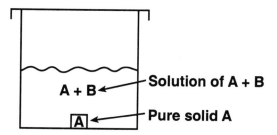

The temperature at which A crystallizes is called the freezing point of the solution.

Critical Thinking Questions

14. Prepare plots of μ vs. T for a pure liquid A and a pure solid A on the same graph. Add a plot of μ vs. T for a solution of B in liquid A to this same graph. Identify the freezing point of pure A and that of the solution. Which is higher?

15. In order for A in the solution to be in equilibrium with pure solid A, what condition must apply to the chemical potentials of A in solution and pure solid A?

16. Write an expression for the chemical potential of A in solution and relate this expression to that for pure solid A at equilibrium.

17. Show that

$$\frac{d \ln X_A(\text{sol})}{dT} = \frac{\Delta_r \overline{H}_{fus}}{RT^2}$$

where $\Delta_r \overline{H}_{fus}$ is the enthalpy of fusion of A. Describe in words the meaning of this relation.

18. If the temperature is increased in a saturated solution of A in equilibrium with solid A, what will you observe? Give a thermodynamic explanation of what you observe.

19. At a fixed temperature describe what happens as a solid C is added to a liquid D to form a saturated solution of C in D.

Exercises

2. Use CTQ 17 and show that $\Delta T_{fp} = K_{fp}m$.

3. The enthalpy of fusion of A refers to the process

$$A(s) \rightleftharpoons A(\ell).$$

The dissolution of $A(\ell)$ in a solution to form an ideal solution is given by

$$A(\ell) \rightleftharpoons A(\text{soln})$$

For the overall process

$$A(s) \rightleftharpoons A(\text{soln})$$

what is $\Delta_r H^\circ$?

4. What is the ideal solubility of $I_2(s)$ at 25°C? $T_{fp} = 113.°C$, $\Delta_r \overline{H}^\circ_{fus} = 15.64\ \text{kJmol}^{-1}$.

5. What is the relation between freezing point depression and solubility?

ChemActivity T19

Osmotic Pressure

Focus Question: A prune is placed in pure water. The inside of a prune is mostly sugar and water. What happens and why?

Model 1: Pure Solvent Separated from a Solution.

Consider an ideal solution of sucrose in water. The solution is separated from the pure solvent at the same temperature by a membrane permeable only to the solvent molecules.

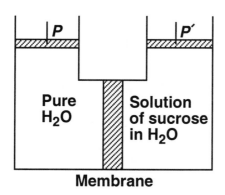

T same in solvent and solution

Water is observed to pass through the membrane.

Critical Thinking Questions

1. Write an expression for the chemical potential of water on both sides of the membrane.

2. On which side of the membrane does water have the lower chemical potential?

3. In which direction does water flow? What is the consequence of this flow?

4. In order for there to be no passage of water molecules through the membrane, what condition on μ must be realized?

5. Recall that $d\overline{G} = -\overline{S}\,dT + \overline{V}dP$. Show that for constant T

$$\mu^*_{H_2O(\ell)}(P) - \mu^*_{H_2O(\ell)}(P') = \overline{V}^*_{H_2O}(P-P')$$

6. Show that $(P-P')\overline{V}^*_{H_2O(\ell)} = RT \ln X_{H_2O(sol)}$.

    Assume that $\overline{V}^*_{H_2O}$ does not depend on P.

7. The osmotic pressure, π, is the excess pressure that must be placed on the solution to prevent diffusion of solvent through the membrane:

$$\pi = P'-P .$$

a) Show that $\pi \overline{V}^*_{H_2O(\ell)} = -RT \ln X_{H_2O(sol)}$.

b) Show that $\pi = -\dfrac{RT}{\overline{V}^*_{H_2O(\ell)}} \ln\left(1-X_{suc(sol)}\right)$

c) Use Taylor's expansion to show that, for dilute solutions,

$$\pi = \dfrac{RT}{\overline{V}^*_{H_2O(\ell)}} X_{suc(sol)}$$

8.   For small $X_{suc(sol)}$ show that

$$\pi = RTC_{suc} \text{ where}$$

C_{suc} is the concentration of sucrose in mol/L.

Exercise

1. Calculate the osmotic pressure of a 1.00 molar solution of sucrose in H_2O at 25°C. To what height would a column of water need to be raised to produce this pressure?

ChemActivity T20

The Phase Rule

Focus Question: Describe a system in which more than two phases exist in equilibrium. Is there a limit on the number of phases that can coexist?

Model 1: Systems of More than One Phase.

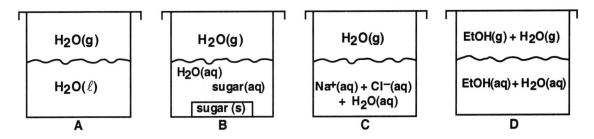

All systems shown are at equilibrium.

Information

Intensive variables are those variables such as temperature, pressure, and mole fraction that are not dependent upon the size of a system.

Extensive variables depend upon the size of the system. Examples are volume, heat capacity, and entropy.

A phase is a region of space in which the macroscopic properties vary continuously as a function of position.

Critical Thinking Questions

1. Identify all four intensive variables in model D above.

2. How many intensive variables are there in A? Identify them.

3. Identify all intensive variables in B and C.

4. How many phases are represented in model A?

5. How many chemically identifiable species are needed to prepare A?

6. In order to completely specify A only one piece of information is needed. What one intensive variable must be known in order to make the system? Why is one variable sufficient?

7. For each model A, B, C, D give the number of phases, p, in the system.

Information

The minimum number of pure chemical substances required for preparation of arbitrary amounts of all phases of the system are called the **components**, c, of the system. This number of pure substances can be found by determining the number of identifiable chemical species and subtracting the number of constraints on the concentration of these species. Equilibrium constants and stoichiometric relationships act as constraints on the concentrations. However, only stoichiometric relationships for which every substance in relationship appears in an equilibrium constant can serve to constrain the number of components.[1]

---

[1] J.S. Alper, *J. Chem. Educ.*, **1999**, *76*, 1567-69.

For example, consider the phase equilibrium for which there are two identifiable species, $H_2O(\ell)$ and $H_2O(g)$

$$H_2O(\ell) \rightleftharpoons H_2O(g)$$

Here, there is one component (H_2O), two species minus one constraint, that of the equilibrium. Thus only one pure chemical substance is needed to prepare the system.

In contrast, consider the dissolution of NaCl(s).

$$NaCl(s) \overset{H_2O}{\rightleftharpoons} Na^+(aq) + Cl^-(aq)$$

In this case, there are two components, NaCl and H_2O. No other pure chemical substances are needed to prepare the system, even though other species ($Na^+(aq)$ and $Cl^-(aq)$) are produced. There are four species, one equilibrium, and the stoichiometric constraint, $[Na^+] = [Cl^-]$.

Critical Thinking Question

8. For each system A, B, C, D give the number of components.

Model 2: The Phase Rule.

When the state of a system can't be completely determined until f intensive variables are given, the system possesses f degrees of freedom. Alternately, the number of intensive variables that can be independently varied without changing the number of phases is called the number of degrees of freedom of the system.

f is given by:

$$f = c - p + 2 \quad \text{Phase Rule}$$

where p is the number of phases.

Critical Thinking Questions

9. State in words the meaning of the above equation.

10. Find the number of degrees of freedom for each system A, B, C, D. For each system list all the intensive variables associated with the system and list the maximum number that must be specified in order to completely determine the system. Provide two examples of the intensive variables which could be specified to determine the system. Explain why there are fewer degrees of freedom than intensive variables.

Exercises

1. Suppose the model system A had been specified as being at 100°C. How many degrees of freedom would the system have?

2. If you were told to prepare a system consisting of pure water in equilibrium with its vapor at 23.8 torr, how would you go about it? How many variables do you need to know to set up this system? What are the possible intensive variables?

3. If you were asked to prepare a solution of sucrose in water, how many intensive variables would you need to know? If the solution is to be saturated, how many intensive variables must be specified?

4. What is the maximum number of phases that can coexist for a one component system? For a two component system? For three components?

5. How many phases are present in the following system? Write the equilibrium constant for the reaction. How many components are there?

$$CaCO_3(s) \rightleftharpoons CaO(s) + CO_2(g)$$

6. How many components are there in the system: $PCl_5(g) \rightleftharpoons PCl_3(g) + Cl_2(g)$? How many phases? If there is no restriction on the ratio of moles of PCl_3, how many components are there?

ChemActivity T21

Solid-Liquid Phase Equilibria

Focus Question: **Water freezes at 0 °C, glycerol at 18 °C. At what temperature would you expect a mixture of these two liquids to freeze?**

Model 1: Cooling Curves.

Heat is removed at a constant rate from samples of H$_2$O and glycerol at 1 bar pressure. The figures below show a plot of temperature vs. time for these processes.

Water

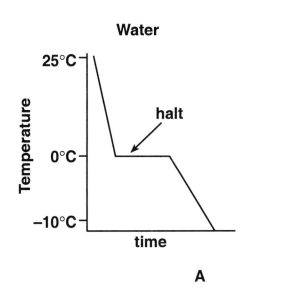

A

Glycerol

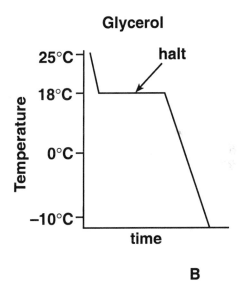

B

Critical Thinking Questions

1. Describe in words how the temperature changes with time as water is cooled from room temperature to –10°C at 1 bar.

2. a) For water between room temperature and 0°C what factor(s) determine the slope of the time-temperature line?

   b) Why is there a halt in the temperature at 0°C and 1 bar?

3. a) For glycerol between 18°C and −10°C what factor(s) determine the slope of the time-temperature line?

   b) Why is there a halt in the temperature at 18°C and 1 bar?

Model 2: Cooling Water and Glycerol.

Ice melts at 0°C and 1 bar, glycerol melts at 18°C and 1 bar. The following behavior is observed when water and glycerol are cooled from room temperature at a pressure of 1 bar.

Pure Water

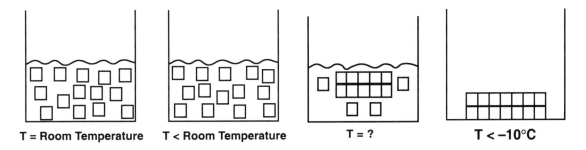

T = Room Temperature   T < Room Temperature   T = ?   T < –10°C

Pure Glycerol

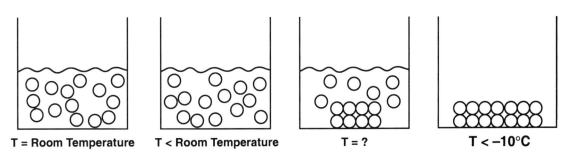

T = Room Temperature   T < Room Temperature   T = ?   T < –10°C

Critical Thinking Questions

4. Describe what happens to glycerol as it is cooled from room temperature to T < –10°C at 1 bar.

5. Only two intensive variables are necessary to completely specify the state of pure water and of pure glycerol at temperatures above 18°C. What are they? If the pressure is fixed at 1 bar, how many intensive variables are necessary?

Model 3: Cooling Curves for Mixtures.

For each mixture of glycerol and water the following time-temperature data were recorded. $P = 1$ bar. For 90% H_2O, some solid first appears at $-10°C$. For 10% H_2O, some solid first appears at 16°C.

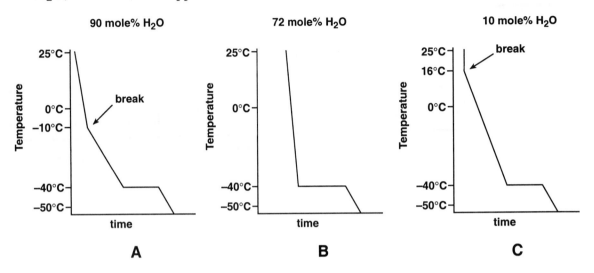

Critical Thinking Questions

6. What does a "break" signify?

7. What is occurring at the "halt" at $-40°C$?

8. For each of the diagrams below, identify which most closely corresponds to each of the following situations: a) 90 mole % water; $-10°C$; b) 72 mole % water, $-40°C$; c) 10 mole % water, 16°C;

$\square$ = water $\bigcirc$ = glycerol $P = 1$ bar

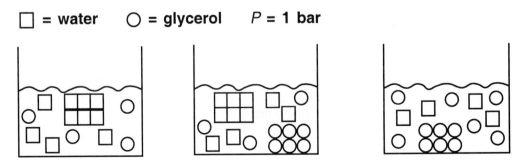

Note: The diagrams <u>do not</u> show the proper ratios of water and glycerol molecules.

Information

For solid-liquid equilibria such as those in CTQ 8, the solid, liquid, and gas phases may be considered. However in general practice if an external pressure greater than the vapor pressure of the liquid phase is applied, the gas phase is said to be cut off. In other words the gas phase is not considered. Thus, in the diagram above on the left at a $P = 1$ bar, only 2 phases are considered to be present.

Critical Thinking Questions

9. How many intensive variables are necessary to completely specify the state of one of these mixtures above 25°C? What are they? If pressure is fixed, how many intensive variables are needed? What are they?

10. How many degrees of freedom does each mixture pictured above have at 1 bar fixed pressure?

Model 4:   Phase Diagram for Water-Glycerol.

The following is a simple eutectic phase diagram for mixtures of water and glycerol.

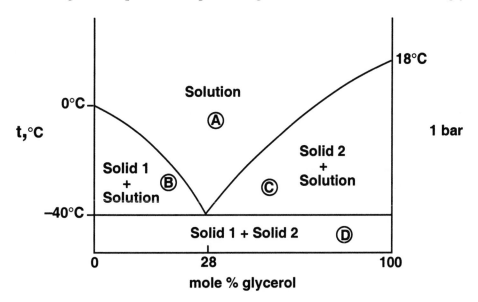

The lowest solution freezing point is called the eutectic point.

Critical Thinking Questions

11. Use the simple eutectic phase diagram to draw cooling curves for mixtures of 80 mole %, 72 mole %, and 25 mole % water.

12. How could a phase diagram such as that above be determined experimentally?

13. Identify solid 1 and solid 2.

14. Identify the phases present at points A, B, C, and D.

15. Give the degrees of freedom at points A, B, C, and D.

Exercises

1. Antimony and lead form a simple eutectic phase diagram. Antimony melts at 630.5°C and 1 bar, lead melts at 327.4°C and 1 bar. A eutectic is observed at 85.5 mole percent lead.

   a) Roughly sketch the phase diagram at 1 bar.

   b) Label each phase region and the eutectic with the number of degrees of freedom.

   c) Sketch a cooling curve for a 10 mole % mixture of lead.

2. Bismuth and cadmium form a simple eutectic phase diagram. Bismuth melts at 271°C and cadmium at 321°C both at 1 bar. The enthalpy of fusion of bismuth is 11.30 kJmol^{-1} and that of cadmium is 6.19 kJmol^{-1}. A eutectic is observed at 55.4 mole percent cadmium.

   a) Calculate the melting temperature of the eutectic mixture. Assume ideal behavior.

   b) Roughly sketch the phase diagram at 1 bar.

   c) Sketch a cooling curve that passes through the eutectic point.

More Complicated Solid-Liquid Phase Equilibria

Focus Question: Is it possible that a mixture of two compounds could freeze at a temperature higher than that of either compound?

Model 1: A Phase Diagram with Compound Formation.

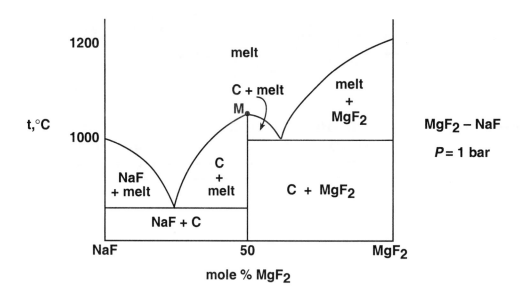

Point M is said to be a congruent melting point. If a solid containing 50 mole % NaF is heated to M, it melts to a liquid of identical composition. A liquid having the composition M exhibits no eutectic halt, behaving like a single pure component.

Critical Thinking Questions

1.   What is the formula of the compound, C, formed?

2.   Draw cooling curves at 90%, 50%, 40%, and 10% MgF$_2$. Identify all halts and breaks.

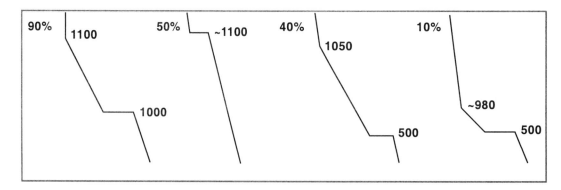

Model 2: Incongruent Melting Points.

When a compound is formed which undergoes decomposition with formation of another solid phase at a temperature below the congruent melting-point of the compound, the equilibrium diagram assumes a different form. An invariant point marking a discontinuity that is not a maximum or minimum point is called a peritectic.

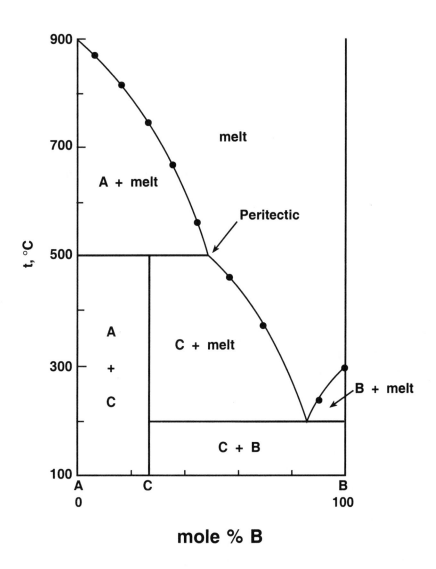

Critical Thinking Questions

3. What is the formula of the compound C formed?

4. Draw cooling curves at 40% B and 10% B.

5. Is there a eutectic point on the diagram? If so, at what composition does it occur?

Exercises

1. The following cooling curves were obtained for a mixture of A and B. Draw the best phase diagram according to these data.

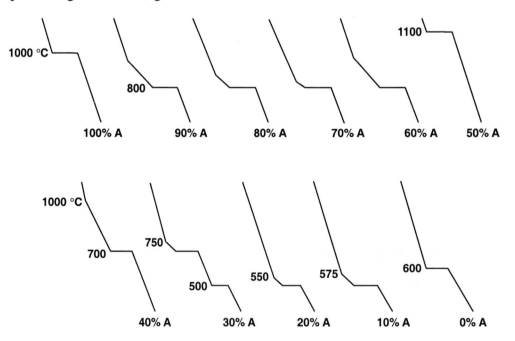

Label the diagram showing what phases are present on each region.

2.

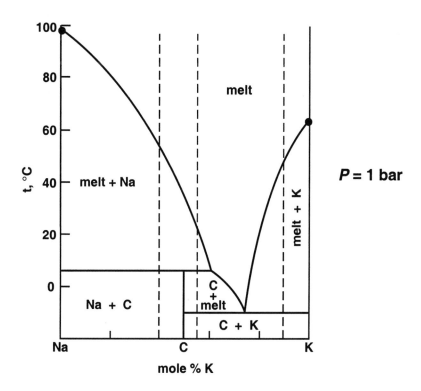

Draw cooling curves along the dotted lines. Identify the compound.

3. Roughly sketch the phase diagram for zinc and magnesium at 1 bar. Zinc melts at
    419°C and magnesium at 650°C. A single compound of 33 mole % Mg is formed and
    melts at 590°C. There are eutectics at 380°C and 347°C, at 10 mole % Mg and 70
    mole % Mg, respectively. Label the phases present and give the number of degrees
    of freedom in each phase region and at each eutectic point.

Liquid-Vapor Phase Equilibria

Focus Question: Is it always possible to separate two liquids by distillation?

Information

At 300 K the vapor pressure of benzene is 103.0 Torr and that of toluene is 32.1 Torr.

Critical Thinking Questions

Note: For many of these CTQs, you may find it useful to use a computer program to create data tables and make the required plots. You may also do these by hand using graph paper.

1. Assume an ideal solution is formed by benzene and toluene at 300 K and complete the following table. The table already includes the data from Table 1 of CA T15.

$X_{bz(sol)}$	P_{bz} (Torr)	P_{tol} (Torr)	P_{tot} (Torr)	$X_{bz(vap)}$
0.100	10.3	28.9	39.2	
0.200	20.6	25.7	46.3	
0.400	41.2	19.2	60.4	
0.600				
0.800	82.4	6.4	88.8	

2. Prepare a plot of the partial pressure of benzene above the solution versus the mole fraction of benzene in the solution. Plot the partial pressure of toluene against the mole fraction of benzene in the solution on the same plot. Add a line for the total pressure as a function of mole fraction of benzene in the solution.

3. Prepare another plot of the total pressure versus the mole fraction of benzene in the solution. Add a plot of the total pressure versus the mole fraction of benzene *in the vapor phase*. Include points for $X_{bz} = 0$ and $X_{bz} = 1$.

Exercises

1. Use the plots made in CTQs 2 and 3 to provide estimates for the following:

   a) For a sample at 300 K, at what pressure does a mixture of one mole each of benzene and toluene begin to boil?

   b) What is the composition of the vapor when boiling occurs at 300 K?

   c) Below what pressure is the system entirely gaseous?

2. Use Raoult's Law and/or Dalton's Law to **calculate** answers to Exercises 1a and 1b.

3. At 100 °C, the vapor pressure of benzene is 1.800 bar and that of toluene is 0.742 bar. What will be the composition of a solution that will boil at 1.000 bar at this temperature?

Model 1: *P,X* Diagram.

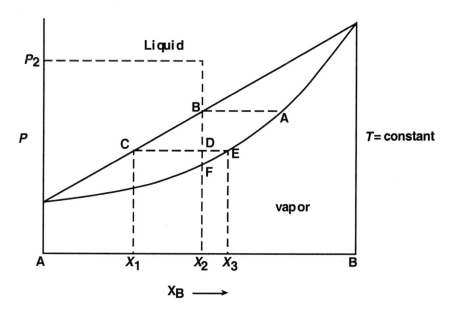

The horizontal line passing through point D (C-E) is called a tie line; it connects the liquid and vapor compositions that are in equilibrium.

The overall composition of the system at point D is X_2. This is made up of a liquid having a composition X_1 and a vapor of composition X_3.

$$X_1 = \frac{n_B^{\ell}}{n_{tot}^{\ell}} \qquad X_3 = \frac{n_B^{v}}{n_{tot}^{v}}$$

where n_B^{ℓ} = moles of B in liquid phase, n_B^{v} = moles B in vapor phase, and n_{tot}^{ℓ}, n_{tot}^{v} are total moles of A and B in liquid and vapor phases.

Critical Thinking Questions

4.   a)   Based on the information in Model 1, explain how the following relationship is obtained:

$$X_2 = \frac{n_B^{\ell} + n_B^{v}}{n_{tot}^{\ell} + n_{tot}^{v}} \qquad (1)$$

b)   It can be shown that

$$\frac{n_{tot}^{\ell}}{n_{tot}^{V}} = \frac{X_3 - X_2}{X_2 - X_1} = \frac{DE}{DC} \qquad (2)$$

This is known as the lever rule and has an important interpretation. What is it?

5.   At point D estimate how much of the system is liquid and how much is vapor.

Model 2: A Plot of Boiling Temperature versus Solution Composition for an Ideal Mixture of A and B.

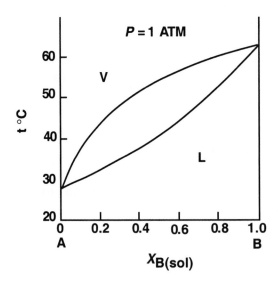

Critical Thinking Questions

6.   What is different about this phase diagram and the one prepared in CTQ 3?

7. Fill in the following data and table from Model 2:

Boiling Point Pure A =

Boiling Point Pure B =

$X_{B(sol)}$	$X_{B(vap)}$	BP (°C)
0.2		
0.4		
0.6		
0.8		

8. Is it possible to start with a 50-50 mixture of A to B and separate the mixture into pure A and pure B? Explain.

Model 3:   Ideal and Non-ideal Solutions.

For two liquids that form an ideal solution a plot of partial pressure vs. mole fraction assumes the form below (see CTQ 2).

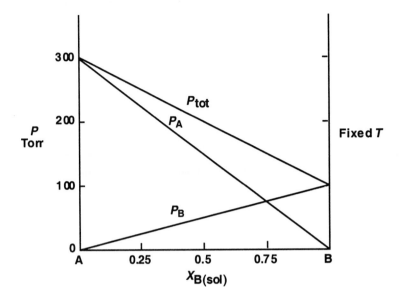

For a non-ideal solution such as that formed by chloroform and acetone, the corresponding plot is given by:

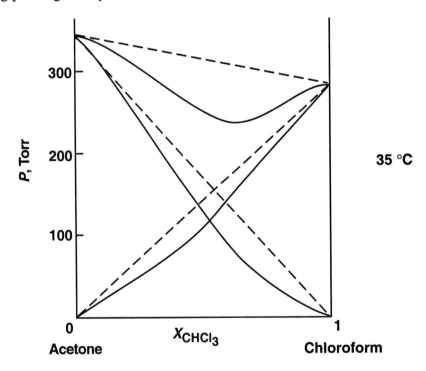

The dotted lines give ideal behavior, the solid lines are actual partial pressures.

Critical Thinking Questions

9. What is the vapor pressure of pure chloroform at 35°C?

10. What is the vapor pressure of pure acetone at 35°C?

11. a) Based on the data given, are the interactions in the mixture of chloroform and acetone stronger than, weaker than, or about the same as the interactions in the pure liquids? Explain your reasoning.

    b) Provide a molecular description to rationalize your answer to part a).

12. This solution is said to exhibit a negative deviation from Raoult's Law. Explain why.

Exercises

4. The lever rule, equation (2), provides an expression for addressing the question: for a given point between C and E in Model 1 (D, for example), what is the relative amount of vapor and liquid? Use the expressions for X, X_2, and X_3 to show that the ratio of the length of the line segments $\dfrac{DE}{DC}$ is equal to the ratio of the number of moles in the liquid and vapor phases $\dfrac{n_{tot}^{\ell}}{n_{tot}^{v}}$. That is, show that $\dfrac{DE}{DC} = \dfrac{n_{tot}^{\ell}}{n_{tot}^{v}}$.

5. Apply the lever rule to find the composition of the vapor and solution when the solution in Exercise 1 exists half in the vapor state and half in the liquid state.

6. Below is a phase diagram for the chloroform-acetone system.

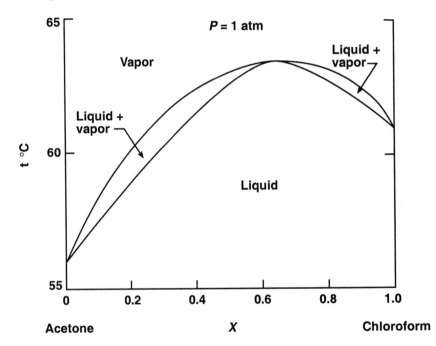

a) What is the composition of the liquid and vapor phases at the maximum boiling point on the diagram?

b) Is it possible to start with a 50-50 mixture of acetone and chloroform and separate the mixture into pure acetone and chloroform? Explain.

c) This type of phase diagram is called a maximum boiling azeotropic diagram. Roughly sketch a minimum boiling azeotropic phase diagram. What interactions between molecules would lead to such behavior? What would a plot of partial pressure vs. solution composition look like? See Model 3. What is meant by a positive deviation from Raoult's Law?

d) What would be the sign of ΔH_{mix} and ΔV_{mix} for minimum and maximum boiling azeotropic mixtures? Explain.

Electrolyte Solutions

Focus Question: **Which solution contains more ionic species 0.1 M NaCl or 0.1 M BaCl$_2$?**

Model 1: Ionic Solutions.

If solutions are not too concentrated, ionic compounds dissociate essentially 100% to form the ions that compose them.

Critical Thinking Questions

1. Which of the following compounds are ionic?

    NaCl, CaF$_2$, Li$_2$SO$_4$, CH$_3$COOH

2. Write an equation that describes the dissolution of the solid ionic compounds in CTQ 1.

3. a) For the general ionic salt, M$_{v_+}$A$_{v_-}$, write an equation that describes its complete dissociation in water. Let Z$_+$ be the valence (charge) of the positive ion and Z$_-$ be the valence of the negative ion.

b) What must be the quantitative relationship involving v_+, v_-, Z_+, Z_- ?

4. For each of the ionic compounds in CTQ 1, give a value for v_+, v_-, and, v, where $v = v_+ + v_-$.

5. a) If μ_+ is the partial molar Gibbs energy of the positively charged ion and μ_- is that of the negatively charged ion, what will be $\Delta_r G$ for the following dissolution process in terms of the Gibbs energy of the solid $\mu^*_{BaCl_2(s)}$, μ_+ and μ_-?

$$BaCl_2(s) \rightleftharpoons Ba^{2+}(aq) + 2Cl^-(aq)$$

b) Provide an analogous expression for the general ionic salt, $M_{v_+}A_{v_-}$.

c) What is the value of $\Delta_r G$ for the saturated solution?

6. From CTQ 5b identify the partial molar Gibbs energy for the ionic part of the electrolyte solution, $\mu_{(ionic)}$.

7. The mean ionic partial molar Gibbs energy, $\mu_{\pm}$, is defined as $\mu_{(ionic)}/v$. Provide an expression for $\mu_{\pm}$ in terms of v_+, v_-, μ_+, μ_-, and v.

Model 2: Partial Molar Gibbs Energy for Solutions.

$$\mu_i = \mu_{i(\ell)}^* + RT\ln X_i \quad \text{(Ideal Solution)} \tag{1}$$

The partial molar Gibbs energy for an ideal solution is given in equation (1), where $\mu_{i(\ell)}^*$ is the molar free energy of pure liquid i and X_i is the mole fraction of component i in the ideal solution. Unfortunately, it is not always appropriate to treat real solutions using equation (1). However, because of the convenience of the form of this equation, an analogous equation for non-ideal solutions is used which includes some correction factors to take into account the non-ideal behavior.

The activity, a_i, can be thought of as the "effective" concentration of component i in a non-ideal solution. The degree of deviation from ideality is generally concentration-dependent, and so the activity can be written as

$$a_i \; = \; \gamma_i m_i \tag{2}$$

where γ_i is the activity coefficient and m_i is the molality. The activity coefficient is defined to be unity at infinite dilution. Thus, for a positively charged ion in solution:

$$\mu_+ = \mu_+^\theta + RT\ln a_+ \tag{2a}$$

$$= \mu_+^\theta + RT\ln \gamma_+ m_+ \tag{2b}$$

where θ is the standard state at infinite dilution.

Critical Thinking Questions

8.   a)   Clearly state the meaning of each symbol in equations (2a) and (2b).

b)   Write an expression analogous to equations (2a) and 2b) for μ_- .

9. a)   Start with your expression for $\mu_\pm$ from CTQ 7 (it may be easier to begin with an expression for $\nu\mu_\pm$) and substitute expressions for μ_+ and μ_- from above. Then collect terms to derive an expression analogous to equation (2a) for $\mu_\pm$,

where $\mu_\pm^\theta = \dfrac{\nu_+\mu_\pm^\theta + \nu_-\mu_-^\theta}{\nu}$ and $a_\pm^\nu = a_+^{\nu+} a_-^{\nu-}$

   b)   Use the expression for $a_\pm^\nu$ from part a) above to show that if $a_\pm = \gamma_\pm m_\pm$, then $\gamma_\pm^\nu = \gamma_+^{\nu+} \gamma_-^{\nu-}$ and $m_\pm^\nu = m_+^{\nu+} m_-^{\nu-}$.

Exercises

1. Using CTQ 9 show that $m_\pm = (\nu_+^{\nu+} \nu_-^{\nu-})^{1/\nu} m$ where m is the molality of the salt.

2. Find $m_\pm$ for each solid ionic salt in CTQ 1 in terms of the molality of the salt.

3. Sketch a pictorial representation of a solution containing a completely dissociated salt like NaCl. Do the same for $BaCl_2$. Sketch a solution containing ions and ion pairs. Define an ion pair and indicate under what circumstances an ion pair might be formed.

4. Why is it almost always necessary to treat electrolyte solutions as non-ideal?

5. The solubilities of the following salts are at a certain temperature.

Ag_2SO_4 0.03 m
PbF_2 1.0 x 10^{-3} m
BaC_2O_4 10^{-4} m

Calculate the concentrations of all ions present in each solution. Identify ν, ν_-, and ν_+. Find $m_\pm$ for each solution.

The Debye-Hückel Theory of Electrolyte Solutions

Focus Question: Which solution is most likely to behave non-ideally: a 0.1 M solution of I_2 in CCl_4 or a 0.1 M solution of NaCl in H_2O? Why?

Model 1: Electrolyte Solution (water molecules not shown).

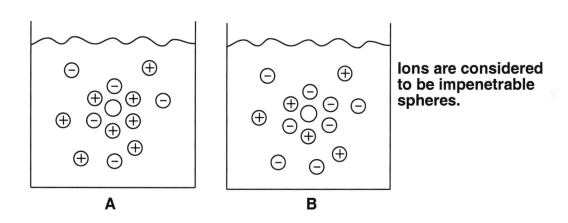

Ions are considered to be impenetrable spheres.

A	B

The electric potential of an isolated charged sphere immersed in a medium of dielectric constant, ε, is

$$V = \frac{Q}{\varepsilon r}$$

(1)

where Q, the charge on the species, is given by $Z_i e$ and V is the electrical potential at a distance r from the sphere. Z_i is the charge, with sign, on the species and e is the absolute value of the charge on the proton.

Let n_+ (and n_-) represent the number of positive (and negative) ions per unit volume around some central sphere. Both n_+ and n_- are dependent on the value of V, so these values vary with distance from the ion.

Critical Thinking Questions

1. Which of the central spheres above is the positive ion and which is the negative ion? Explain the rationale for your choice.

2. a) Is the potential V surrounding a positively charged sphere positive or negative?

    b) Is the potential V surrounding a negatively charged sphere positive or negative?

    c) What happens to the potential (for positive and negative species) as r approaches infinity?

    d) For a positively charged central sphere, do you expect n_+ to be larger near the central sphere (small r) or further from the sphere (larger r)? Explain your reasoning.

3. a) Assuming n_+^0 is a constant, which of the two mathematical relationships below best describes the number of positive ions per unit volume around a central positive sphere? Explain your analysis clearly.

    $$n_+ = n_+^0 \, e^{-eZ_+ V/kT} \qquad\qquad n_+ = n_+^0 \, e^{eZ_+ V/kT}$$

b)   What is the interpretation of n_+^0?

c)   Answer CTQ 3a for the number of negative ions per unit volume n_- around a central positive ion. Explain your analysis clearly.

d)   For a simple univalent salt (such as NaCl), how does n_+^0 compare to n_-^0? Explain your reasoning.

Information

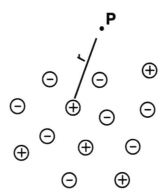

The electrical potential at a point P at a distance r from a reference ion is determined by the potential of the reference ion and the potential due to the ionic atmosphere produced by all the other ions.

The potential at P as a function of r can be found from the Poisson equation:

$$\frac{d^2V}{dr^2} + \frac{2}{r}\frac{dV}{dr} = -\frac{4\pi\rho}{\varepsilon} \tag{3}$$

where ρ is the charge density, the charge per unit volume, and ε is the dielectric constant for the medium.

Critical Thinking Questions

4. For a central positive ion, we can construct an expression for the charge density, ρ, as a sum of contributions from both the positively and negatively charged ions.

    a) Provide an expression for ρ around a central positive ion in terms of n_+, n_-, Z_+, Z_-, and e.

    b) Substitute your answers from CTQ 3 for n_+ and n_- in your expression above to obtain an expression for ρ showing its dependence on V (and other variables).

5. Expanding the two exponential terms of ρ for a central positive ion in a Taylor's series, it can be shown that

    $$\rho = e\left\{ n_+^0\, Z_+ + n_-^0\, Z_- - \frac{eV}{kT}\left(n_+^0\, Z_+^2 + n_-^0\, Z_-^2 \right) \right\} \tag{4a}$$

    $$\rho = -\frac{e^2 V}{kT}\left(n_+^0\, Z_+^2 + n_-^0\, Z_-^2 \right) \tag{4b}$$

    a) Why does the $\left(n_+^0\, Z_+ + n_-^0\, Z_- \right)$ term vanish when eq. (4a) is simplified to give eq. (4b)?

    b) Based on eq. 4, for a simple univalent salt (such as NaCl), which of the following best describes the charge density surrounding a central positive ion: always positive, always negative, sign varies with distance, cannot determine? Explain.

Information

The ionic strength is given by

$$I = 1/2 \left(n_+^0 Z_+^2 + n_-^0 Z_-^2 \right) \tag{5}$$

Commonly, the ionic strength is defined in terms of molalities: $I = 1/2 \sum_i m_i Z_i^2$.

Critical Thinking Question

6. Use equation (5) to show that

$$\rho = -\frac{2e^2 V I}{kT}$$

Information

Now that we have an expression for ρ, equation (3) can be solved. It can be shown that

$$V = \frac{Ce^{-\kappa r}}{r} \tag{6}$$

where C is a constant equal to $\frac{Z_+ e}{\varepsilon}$ and $\kappa^2 = \frac{8\pi e^2 I}{\varepsilon kT}$

Critical Thinking Question

7. Expand the exponential equation (6) in a Taylor's series to show

$$V = \frac{C}{r} - C\kappa ,$$

and then make appropriate substitutions to show that $V = \frac{Z_+ e}{\varepsilon r} - \frac{Z_+ e\kappa}{\varepsilon}$

Information

Electrical work is potential acting against charge, i.e., VQ or VZ_ie. VQ is the work required to bring ions i and j from infinite separation to the required configuration in the presence of all the other ions and solvent.

The work required to positively charge a central neutral atom in the presence of all the other ions is:

$$w_+ = \int_0^Q V \, dQ = \int_0^{Z_+e} V d(Z_+e) \tag{7}$$

Critical Thinking Questions

8. Integrate the right-hand side of equation (7), substituting the expression for V from CTQ 7, to obtain an expression for w_+ in terms of Z_+e.

9. The potential of a charged sphere in a medium with a dielectric constant ε but with no other ions present is as given in Model 1,

$$V = \frac{Q}{\varepsilon r} = \frac{Z_ie}{\varepsilon r} \, .$$

a)   Find the work to charge this sphere from zero to Q.

b)   Note that the answer to CTQ 9a is identical to one of the terms from CTQ 8. Use grammatically correct English sentences to identify the meaning of each of the two terms from CTQ 8.

Model 2: The Debye-Hückel Model.

The Debye-Hückel model for an electrolyte solution is that if there were no interactions due to the charges on the ions, the solution would behave ideally. That is, the chemical potential would be given by:

$$\mu_i = \mu_i^\theta + RT \ln m_i \, . \tag{8}$$

Critical Thinking Questions

10. Recall that $\mu_i = \mu_i^\theta + RT\ln a_i$. Make appropriate substitutions to generate an expression in terms of m_i and γ_i. Be sure to separate terms to clearly show how this expression deviates from that in equation (8).

11. According to D-H electrolyte theory, any difference in μ_i from that given in equation (8) is equal to the work required to charge the ion from zero to Q in the presence of all the other ions. Note that we have derived two expressions for this quantity: one arose in CTQ 10 and the other in CTQ 8. (If you are unsure about this, examine your answer to CTQ 9b.)

    a) Provide an expression relating γ_i (or $RT \ln \gamma_+$) to Z_+, e, and κ.

    b) Show that $\ln\gamma_+ = -AZ_+^2 I^{1/2}$ where A is a collection of constants at a given temperature.

    c) According to the D-H electrolyte theory, how should the activity coefficient γ_+ change as the temperature is increased (assuming constant volume)? Explain your reasoning carefully.

Exercises

1. Expand the two exponential terms in your answer to CTQ 5 to obtain equation (4) for a central positive ion.

2. Actually there is another solution to Poisson's equation,

$$v = \frac{Ce^{-Kr}}{r} + \frac{Be^{Kr}}{r}$$

where B is a constant and C is as previously defined. Why is this not an acceptable solution? Hint: Consider what happens to v as r gets very large.

3. a)   Derive the relationship $\ln\gamma_- = -AZ_-^2 \, I^{1/2}$.

    b)   Use the relationship between $\gamma_\pm$, γ_+, and γ_- to show that

$$\ln\gamma_\pm = AZ_+ Z_- I^{1/2}$$

    where A is a constant.

4. Calculate the ionic strength of the following solutions: 0.050 m NaCl, 0.050 m CaCl$_2$, 0.050 m MgSO$_4$.

ChemActivity E3

The Structure of Electrolyte Solutions

Focus Question: Is $BaSO_4$ more soluble in water or 0.1 M KNO_3?

Information

It is possible to determine mean ionic activity coefficients for ions in solution experimentally. Solubility studies, conductance measurements, and electrochemical cells all have allowed the calculation of mean ionic activities and activity coefficients.

The ionic strength is defined in terms of molality: $I = \frac{1}{2} \sum_i m_i Z_i^2$.

Model 1: Mean Activity Coefficients as a Function of Ionic Strength.

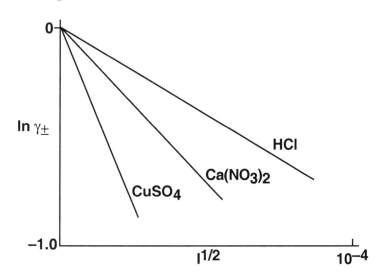

For dilute solutions of the above salts, the slopes of the plots were experimentally found to be:

	HCl	Ca(NO₃)₂	CuSO₄
Slope	−1.17	−2.34	−4.68

Critical Thinking Questions

1. What is the behavior of $\ln \gamma_\pm$ versus $I^{1/2}$ over the region shown in the plot? Write a general equation that expresses this relationship.

2. What would be the slope of $\ln \gamma_\pm$ vs $I^{1/2}$ plot for Ag_2SO_4 in dilute solution? What is the general relationship for the slope of $\ln \gamma_\pm$ vs $I^{1/2}$ for an electrolyte? Give an expression for $\ln \gamma_\pm$.

3. Why is the slope of the line negative? Hint: To what thermodynamic function is $\gamma_\pm$ directly related?

4. If the central ion were not surrounded by other ions, what would be $\ln \gamma_\pm$? Why is $\ln \gamma_\pm = 0$ the common origin for these plots?

Exercises

1. (a) Calculate the mean ionic activity coefficient for $RbCl_2$ in a $0.00500\ m$ solution at $25°C$.

2. Calculate the mean ionic activity coefficient for a $1.0 \times 10^{-4}\ m$ and for a $1.0 \times 10^{-3}\ m$ solution of HCl at $25°C$.

   Do the same calculation for the same molalities of a $CuSO_4$ solution.

   Explain any differences found for $\gamma_\pm$ in the two solutions.

Model 2: $\gamma_\pm$ as a Function of Ionic Strength.

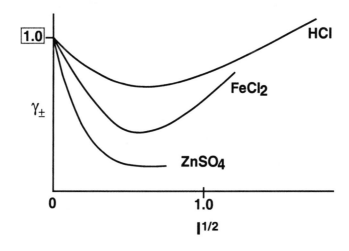

For some real ionic compounds, at high ionic strength the water molecules are "tied up" so that the effective concentration of the ions increases. In other cases, ion pairs form in solution, so that increasing the amount of dissolved material does not result in actually increasing the concentration of free ions.

Critical Thinking Questions

5. Place a value on the tick mark along the y-axis in the plot. Explain why all plots begin at a common origin.

6. Are the plots shown in Model 2 consistent with the model developed in CTQ 2? Explain your reasoning.

7. For each ionic compound presented in Model 2, determine whether or not ion pair formation is observed at high ionic strength. Explain your reasoning.

Model 3:   Balancing Ion Distribution Function.

A general plot of the probability P(r) of finding a balancing ion (an ion of opposite charge) at a distance r from the central reference ion is given below.

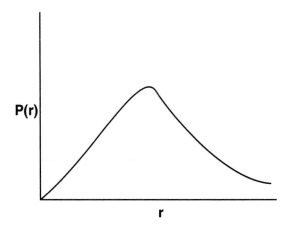

Critical Thinking Questions

8. What information does the above plot convey? Roughly sketch a picture of an electrolyte solution consistent with this plot.

9. The maximum on the plot above occurs at $1/\kappa$ (see CA E2). The potential felt by a reference ion due to all the other ions is $V = -\dfrac{Z_+e}{\varepsilon\,(1/\kappa)}$, as shown previously (CA E2). Interpret what these two observations mean.

Information

$$\kappa^2 = \frac{8\pi e^2}{\varepsilon kT} \, I$$

Critical Thinking Questions

10. The collection of constants above can be calculated at 25°C to be 10.76×10^{18} Kg mol^{-1}m^{-2} when I has units of mol Kg^{-1}. Find $1/\kappa$. Interpret what $1/\kappa$ means for solutions of 0.001 m, 0.01 m, and 0.1 m HCl.

Model 4: A Weak Acid, HA, is not Completely Dissociated in Aqueous Solution.

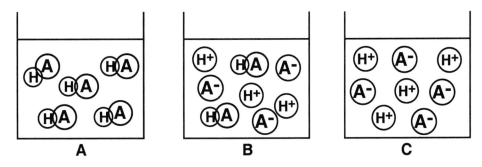

The degree of dissociation, α, of a weak acid is defined as $\alpha = \frac{x}{m_0}$ where x represents the molality of dissociated acid and m_0 is the initial molality of the acid before dissociation.

Critical Thinking Questions

11. Which of the above representations best represents a weak acid? Which best represents a strong acid? Explain.

12. Write the chemical equation that describes the dissociation of the weak acid CH_3COOH.

13. Write the equilibrium constant, K_a, for the above reaction in terms of the mean ionic activity coefficient and the molalities of the various components. Explain why the activity coefficient γ_{HA} for the weak acid can be omitted from this expression.

14.   a)   For a weak acid, is α large or small?

   b)   Show that $\alpha = \dfrac{1}{\gamma_{\pm}}\sqrt{\dfrac{K_a}{m_O}}$ for a weak acid. State clearly any assumptions or approximation made.

   c)   At relatively low ionic strength, what happens to the degree of dissociation, α, as the ionic strength of the medium increases? Give a mathematical and a physical explanation.

15. Describe how the degree of dissociation could be made to increase.

Exercises

3. A 0.0104 *m* chloroacetic acid solution is 28.0% dissociated at 25°C. Calculate K for the reaction of chloroacetic acid with water neglecting any non-ideal behavior. Calculate K by taking into account the non-ideal behavior.

4. The equilibrium constant for the dissociation of acetic acid is 1.75×10^{-5} at 25°C. Calculate the concentration of H_3O^+ in an acid solution of initial concentration 0.10 *m*.

Information

Many salts are relatively insoluble in water. If a saturated solution of an ionic compound is greater than about 0.1 *m*, the compound is said to be soluble. If the saturated solution is less than about 10^{-3} *m* the compound is said to be insoluble. Intermediate cases are moderately soluble.

Model 5: K_{sp} for $BaSO_4$.

K_{sp} for $BaSO_4$ is 1.05×10^{-10} at 25 °C.

Critical Thinking Questions

16. Make a rough sketch of a saturated solution of barium sulfate. Identify all species present.

17. Write the equilibrium constant, K_{sp}, for the dissolution of barium sulfate in terms of molalities and activities. Give the chemical equation that describes the dissolution. Remember to treat ions properly.

18. If S represents the molal solubility of the $BaSO_4$, what are the ion concentrations in terms of S?

19. Sketch a plot of $\ln S$ versus $I^{1/2}$ for $BaSO_4$.

Exercises

5. Show that for a 1:1 ionic compound that forms a saturated solution in water $\gamma_\pm = \dfrac{S_o}{S}$ where S_o is the solubility at zero ionic strength.

6. What would happen to the solubility of $BaSO_4$ if KNO_3 were added to a saturated solution?

7. The solubility of PbI_2 in water at 20°C is 1.37×10^{-3} m. A = 1.167 at 20°C.

    a) Calculate the K_{sp} for PbI_2.

    b) Calculate the solubility of PbI_2 in 0.030 m KI.

    c) Find the solubility of PbI_2 in 0.030 m KNO_3.

    d) Ignore any electrical interaction effects; i.e., assume the value that $\gamma_\pm$ would have if the solution were ideal and calculate the solubility of PbI_2. By what percentage would this calculation be in error?

Electrode Potentials

Focus Question: In an electrochemical cell one electrode is negative. Why?

Model 1: $dG \leq dw_{\text{non-}PV}$ T,P.

An example of non-PV work is electrical work.

$$\nu_A \qquad\qquad\qquad \nu_B$$

Points A and B represent spheres of two identical metal phases each with a different electrical potential, ν_i imposed upon it.

Critical Thinking Questions

1. Write the total derivative for $G = G(n_1, n_2)$ T,P constant. Substitute the appropriate symbol for the partial molar Gibbs energy.

2. If dn electrons are reversibly transferred from Sphere B at potential ν_B to Sphere A at potential ν_A, show how to obtain the following for the net transfer.

$$dw_{e\ell\text{net}} = dG = \bar{\mu}_{eA}dn - \bar{\mu}_{eB}dn$$

$\bar{\mu}_{ei}$ represents the Gibbs energy of the electrons in phase i.

3. Electrical work is defined as moving a charge, Q, through a potential. Rationalize that for the net transfer of electrons in CTQ 2:

$$dw_{el_{net}} = V_A dQ - V_B dQ = EdQ$$

Define E.

Information

The amount of charge, Q_i, on a charged species is given by $Q_i = Z_i F n_i$ where F is Faraday's constant, the charge on an Avogadro's number of electrons, 96485 $Cmol^{-1}$.

Critical Thinking Questions

4. Show that

$$dw_{el} = V F Z_i dn_i$$

for the movement of dn_i moles of electrons into or out of a potential field.

For the net transfer of electrons from point B to A find $dw_{el_{net}}$ in terms of charges and potentials.

5. Show that the partial Gibbs energy of the electrons transferred to point A is

    $$\bar{\mu}_{eA} = Z_- F V_A.$$

    What would be a similar expression for $\bar{\mu}_{eB}$?

6. Generalize these results to any species with charge Z_i under any potential V_i.

7. If the electrical potential imposed on a phase is negative, what would be the sign of the partial Gibbs energy for electrons in that phase? What would be the sign of the partial Gibbs energy for positive ions in that same phase? What does this mean?

Model 2: The Hydrogen Electrode.

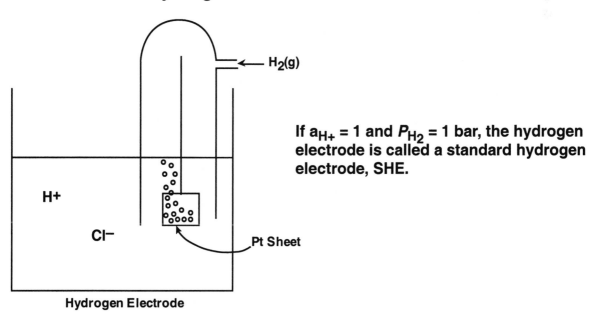

$\longleftarrow H_2(g)$

If $a_{H+} = 1$ and $P_{H_2} = 1$ bar, the hydrogen electrode is called a standard hydrogen electrode, SHE.

H+

Cl−

Pt Sheet

Hydrogen Electrode

Critical Thinking Questions

8. In the hydrogen electrode H^+ migrates to the Pt sheet. What reaction takes place at the Pt surface? Clearly label all states. Indicate the phase in which the electrons are found. By convention chemists write electrode reactions as reduction reactions. Make sure your electrode reaction conforms to this convention.

9. Give the condition for chemical equilibrium using partial Gibbs energies for all species.

10. Substitute for each partial Gibbs energy in terms of μ_i^θ, activities, and potentials and solve for the potential developed at the Pt surface.

Information

By convention $\mu_{H^+(aq)}^\theta$ is taken to be zero and V_{SHE}^θ is thus zero. A single electrode as shown in Model 2 is called a half-cell.

Model 3: A Metal-Metal Ion Electrode.

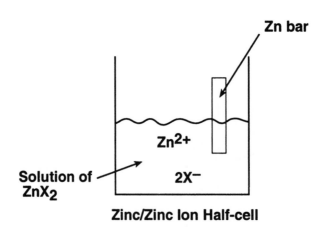

Zinc/Zinc Ion Half-cell

Critical Thinking Questions

11. By analogy with Model 2 find an expression for the electrode potential of the $Zn(s)/Zn^{2+}(aq)$ electrode. Identify the standard electrode potential, V^{θ}, and define V^{θ}. Let $\mu^{*}_{Zn(s)}$ represent the standard Gibbs energy of solid zinc.

12. What is the electrode potential, V, for the standard hydrogen electrode?

13. If the zinc half-cell above is connected to a standard hydrogen electrode show how E_{cell} is related to the zinc and SHE electrode potentials.

14. Why are electrode potentials always measured relative to the SHE?

Exercises

1. A metal-metal ion electrode consists of a bar of metal in a solution of the metal ion.

$$M^{n+}(aq) + ne^{-} = M(s)$$

Give the expression for this electrode potential.

2. Find an expression for the electrode potential for

$$Cl_2(g) + 2e^{-} = 2Cl^{-}(aq)$$

3. Write the electrode potential for

$$Fe^{3+}(aq) + e^- = Fe^{2+}(aq)$$

4. Define the state of the electrons in each reaction above.

Information

Representation of half-cells. The half-cells such as those described above are conventionally represented by a diagram that shows the oxidized and reduced forms separated by a solid vertical bar whenever there is a phase separation. Commas separate different species in the same phase.

Critical Thinking Questions

15. The conventional representation for the hydrogen electrode is

$$H^+(m) \mid H_2(g) \mid Pt(s).$$

Identify each symbol and bar.

16. Give the conventional representation for the metal-metal ion electrode.

Exercises

5. Roughly sketch the half-cell to which the following representation refers.

$$Cl^- (0.005 \ m) \mid Cl_2(1 \ bar) \mid graphite$$

6. Give the representation for the half-cell in which the following reaction occurs.

$$Sn^{4+}(aq) + 2e^-(Pt) = Sn^{2+}(aq)$$

7. Sketch the half-cell to which the following representation refers:

$$Cl^-(1 \ m) \mid AgCl(s) \mid Ag(s)$$

Electrochemical Cells

Focus Question: **Why does a battery powered watch die all at once?**

Information

Two half-cells may be combined to produce an electrochemical cell. If two half-cells require a connection known as a salt bridge, double vertical bars are used to indicate the junction. The direction of electron flow in the external circuit is from negative to positive.

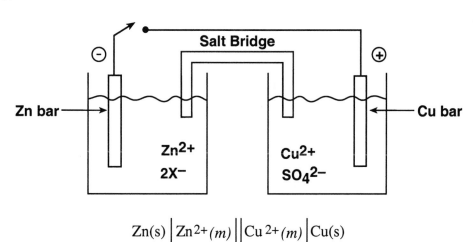

$$\text{Zn(s)} \left| \text{Zn}^{2+}(m) \right| \left| \text{Cu}^{2+}(m) \right| \text{Cu(s)}$$

Critical Thinking Questions

1. Roughly sketch the cell given by the representation:

$$Pt(s) \mid H_2(g) \mid H^+(m) \parallel Zn^{2+}(m) \mid Zn(s)$$

2. Give the representation for the electrochemical cell

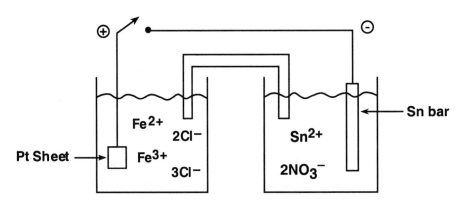

Write the half-reactions for each cell.

Conventional Representation

An electrochemical cell has two electrodes. The potential of the cell, E, is defined by

$$E = V_{RIGHT} - V_{LEFT}$$

where V_{RIGHT} and V_{LEFT} are the electrode potentials of the half-cells. The choice as to which half-cell is designated as the right-hand cell and which is the left-hand cell is arbitrary. Once chosen the half-cell designations cannot be reversed.

Model 1: An Electrochemical Cell.

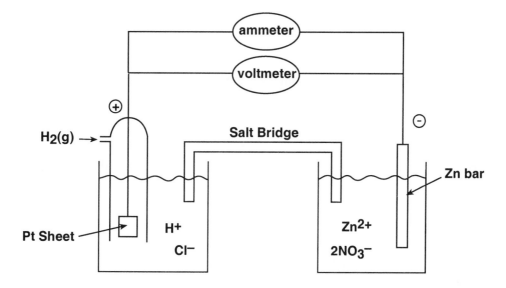

Critical Thinking Questions

3. Arbitrarily select the Zn/Zn^{2+} electrode to be the right-hand electrode. Write the expression for E in terms of the electrode potentials of the half-cells.

4. Write chemical equations that describe reduction reactions at the right and left electrodes. Be sure to indicate in what phases the electrons are to be found.

5. Subtract the equilibria in CTQ 4 in the same sense as the potentials, right minus left. This resulting chemical equation summarizes the electrical and chemical changes that occur in the cell.

6. Write the condition on the partial Gibbs energy necessary for equilibrium for the reaction in CTQ 5.

7. Collect the partial Gibbs energy terms into chemical and electrical components.

8. Identify $\Delta_r G$, the Gibbs energy change for the chemical reaction.

9. Substitute electrode potential expressions for the partial Gibbs energy for the electrons.

10. Show $\Delta_r G = -2F\text{E}$ where E is $(V_{Zn2+,Zn} - V_{H+,H_2})$

11. Generalize CTQ 10 to apply to any electrochemical cell. Let n represent the number of moles of electrons transferred per mole of reaction.

12. Complete the following table.

$\Delta_r G$	E	Cell reaction is
–	+	
		non-spontaneous
0	0	

Model 2: $\Delta_r G = \Delta_r G^\theta + RT \ln Q$.

Critical Thinking Questions

13. State in words what the above equation means.

14. Show that

$$E = E^\theta - \frac{RT}{nF} \ln Q \qquad \text{Nernst Equation}$$

Identify each symbol.

15. For the cell

$$Pt(s)| H_2(g) | H^+(m) || Zn^{2+}(m) | Zn(s)$$

Write the electrode potential expression, V_i, for each half-cell.

16. Following an earlier choice, let the zinc-zinc ion electrode be designated as the right-hand electrode and subtract the two electrode potentials.

Write the overall cell reaction for the net chemical reaction.

17. Show how CTQ 16 relates to the Nernst equation for this cell.

18. Use the three electrochemical cells in this ChemActivity and a table of standard electrode potentials to answer the following question.

Is it true that the electrode with the most positive value of V^θ will generally be the positive electrode? Explain.

Information

For the cell of CTQ 15, E is –0.76 V when all species are in their standard states. The hydrogen cell is therefore a Standard Hydrogen Electrode, SHE.

Critical Thinking Questions

19. Write the overall cell reaction in the spontaneous direction.

20. Write the half-cell reaction that occurs in the SHE when the reaction proceeds spontaneously.

21. Subtract the half-reaction in CTQ 20 from the overall reaction in CTQ 19 and show that

$$Zn^{2+}(aq) + 2e^-(Pt) = Zn(s) \ .$$

Information

In CTQ 21 note that the equation is a shorthand way of summarizing the overall process:

$$Zn(s) + 2H^+(a=1) = Zn^{2+}(aq) + H_2(P=1bar)$$

where

$$2H^+(a=1) + 2e^-(Pt) = H_2(P=1bar)$$

is subtracted from the overall reaction. Thus electrode reduction potentials measured against the SHE are written as

$$Zn^{2+}(aq) + 2e^-(Pt) = Zn(s)$$

or more commonly

$$Zn^{2+}(aq) + 2e^- = Zn(s)$$

The electrons that appear in the reduction potential cell reaction thus show that the potential has been obtained by measurement against a SHE.

Exercises (Note: For some of these exercises a table of standard reduction potentials will need to be consulted.)

1. For the cell

$$Zn(s) \ \Big| \ Zn^{2+}{}_{(m)} \ \Big|\Big| \ Cu^{2+}{}_{(m)} \ \Big| \ Cu(s)$$

give the electrode potential expression, V, for each half-cell. Write the Nernst equation.

$$V^\theta(V)$$

	$V^\theta(V)$
$Cu^{2+}(aq) + 2e^- = Cu(s)$	+0.337
$Zn^{2+}(aq) + 2e^- = Zn(s)$	−0.763

Calculate the equilibrium constant for the spontaneous cell reaction at 25°C.

Roughly sketch the cell and show the direction of electron flow in the external circuit.

2.   Calculate K_{sp} for AgBr at 25°C.

$$AgBr(s) = Ag^+(aq) + Br^-(aq)$$

	$V^\theta(V)$
$AgBr(s) + e^- = Ag(s) + Br^-(aq)$	0.0713
$Ag^+(aq) + e^- = Ag(s)$	0.7991

3.   The potential of the cell

$$\ominus Zn(s) \mid ZnCl_2(m = 0.01021) \mid AgCl(s) \mid Ag(s) \oplus$$

is $1.1566V$ at 25°C. Find the mean ionic activity coefficient for $ZnCl_2$.

4.   For the cell

$$SHE \mid Ag^+(m) \mid Ag(s)$$

show that

$$E - 0.05910 \log m = E^\theta_{Ag^+,Ag} - 0.029\ m^{1/2}$$

How could $E^\theta_{Ag^+,Ag}$ be determined? SHE represents the standard hydrogen electrode.

5.   For the cell

$$Pt(s) \mid H_2(P=1) \mid H^+(m), Cl^-(m) \mid AgCl(s) \mid Ag(s)$$

show that

$$E = E^\theta - 2\frac{RT}{F} \ln m - 2\frac{RT}{F} \ln \gamma_\pm \ .$$

Roughly sketch the cell and indicate which is the positive and which is the negative electrode. What is the spontaneous cell reaction?

6.   The cathode is the electrode at which reduction occurs for the overall cell reaction. For the first three cells diagramed in this ChemActivity, identify the cathode and the anode.

7.   What is the standard partial Gibbs energy for $Zn^{2+}(aq)$ compared to that for $H^+(aq)$? Why is the zinc electrode negative when connected to a SHE? What does this mean about any negative electrode in a cell?

8.   Would the cell

$$Pt(s) \mid H_2(1\ bar) \mid HCl(m_1) \parallel HCl(m_2) \mid H_2(1\ bar) \mid Pt(s)$$

develop a potential? Explain why or why not.

Temperature Dependence
of Cell Voltage

Focus Question: Why does the voltage of a cell depend on temperature?

Model: $\Delta_r G = -nF\mathrm{E}$.

Critical Thinking Questions

1. State in words what the above equation means.

2. Show that

$$\left.\frac{\partial \mathrm{E}}{\partial T}\right)_P = \frac{\Delta_r S}{nF}.$$

3. Show that

$$\Delta_r H = -nF\mathrm{E} + nFT \left.\frac{\partial \mathrm{E}}{\partial T}\right)_P$$

4. Find an expression for $\Delta_r G^\theta$ in terms of standard cell voltage. What are $\left.\dfrac{\partial E^\theta}{\partial T}\right)_P$ and $\Delta_r H^\theta$?

Exercises

1. Find the temperature dependence of E^θ for:

$$Pt(s)\,|\,H_2(P{=}1)\,|\,H^+(aq)\,||\,OH^-(aq)\,|\,O_2(g)\,|\,Pt(s)$$

$$V^\theta(v)$$

$O_2(g) + 2H_2O(\ell) + 4e^- = 4OH^-(aq)$ ⠀⠀⠀⠀ 0.401

$H^+(aq) + e^- = 1/2\ H_2(g)$

	$\Delta_r G^\theta$(kJmol^{-1})	$\Delta_r H^\theta$(kJmol^{-1})
$H_2(g) + 1/2\ O_2(g) = H_2O(\ell)$	-237	-285
$H_2O(\ell) = H^+(aq) + OH^-(aq)$	79.9	56.9

2. For the cell in which the reaction is

$$Hg_2Cl_2(s) + H_2(1\ bar) = 2Hg(\ell) + 2HCl(aq)$$

$E^\theta_{298} = +0.2676\ V$ and $\left.\dfrac{\partial E^\theta}{\partial T}\right)_P = -3.19 \times 10^{-4}\ V/K.$

Find $\Delta_r G^\theta$, $\Delta_r H^\theta$, and $\Delta_r S^\theta$.

Introduction to Chemical Kinetics

(What Is Kinetics?)

Focus Question: **In a rate law for a chemical reaction, how many (and which) of the following must be true about the order of a reaction with respect to each reactant?**

a) equal to the stoichiometric coefficient

b) an integer

c) non-zero

d) positive

Information

When a chemical reaction occurs, reactants are consumed and products are produced. By convention, rates of consumption and rates of production of chemical species are always reported as positive numbers:

$$\text{rate of consumption of reactant} = -\frac{\Delta[\text{reactant}]}{\Delta\text{time}} \qquad (1)$$

Model 1: A Chemical Reaction.

When ClO^- is added to water, a chemical reaction occurs. Table 1 displays some data relating to such an experiment.

Table 1. Experimental data for a chemical reaction

Time (s)	Molarity of ClO^- (M)	Molarity of Cl^- (M)	Molarity of ClO_3^- (M)
0	2.40	0	0
1.00×10^2	1.80	0.40	0.20

R.S. Moog and J.J. Farrell, *Chemistry: A Guided Inquiry*, 2nd Ed., John Wiley & Sons, Inc., 2nd Ed., 2002, CA 36. This material is used by permission of John Wiley & Sons, Inc.

Critical Thinking Questions

1. Why is there a negative sign in equation (1)?

2. Provide an expression for the rate of production of a product analogous to equation (1).

3. Based on the data in Table 1, what is the rate of production or consumption, in M/s, based on the first 100 seconds, for each of the three components listed?

4. Based on your answer to CTQ 3, is it obvious what the rate of reaction is for the process taking place? Explain your reasoning.

Information

A balanced chemical equation may be written as

$$a\text{A} + b\text{B} \rightarrow c\text{C} + d\text{D} \tag{2}$$

Let $n_{i,o}$ be the number of moles of component i present in a container initially.

Let ξ be defined as the **extent** of the reaction. After ξ moles of reaction have occurred, there will be $n_{\text{A},o} - a\xi$ moles of A in the container.

The **rate of reaction** is defined to be

$$\text{rate of reaction} = \frac{d\xi}{dt} \tag{3}$$

Critical Thinking Questions

5. How many moles of each of the components, (A, B, C, D) are present in the container after ξ moles of reaction have occurred?

6. Use your answer from CTQ 5 to obtain an expression for $\dfrac{dn_A}{dt}$ in terms of a and $\dfrac{d\xi}{dt}$.

7. Use your answer from CTQ 6 to obtain an expression for $\dfrac{d\xi}{dt}$ in terms of a and $\dfrac{dn_A}{dt}$.

Exercises

1. Consider the reaction of hydrogen gas and oxygen gas to form water. An experimental investigation of a reaction vessel in which this reaction was occurring showed that the rate of loss of hydrogen at some time was $1.64 \times 10^{-5} \dfrac{\text{moles}}{\text{hour}}$ under certain conditions.

    a)   What is the rate of loss of oxygen in this vessel under these conditions?

    b)   What is the rate of reaction under these conditions?

2. Indicate whether the following statement is true or false and explain your reasoning:

    The rate of a reaction is equal to the rate of consumption of each of the reactants.

3. Consider the reaction taking place in Model 1.

    a)   Provide a balanced chemical equation for the reaction.

    b)   Assuming that the reaction is taking place in a 1.00-liter solution, determine the rate of reaction, in M/s, based on the first 100 seconds.

Information

As we have seen, at any time in a reaction,

$$n_i = n_{i,0} + v_i \xi \tag{4}$$

where is v_i is the stoichiometric number of i. The stoichiometric numbers, which are dimensionless, are positive for products and the negative for reactants. For example, for the reaction

$$H_2 + \frac{1}{2}O_2 \rightarrow H_2O$$

$$v_{H_2} = -1 \qquad v_{O_2} = -1/2 \quad \text{and} \quad v_{H_2O} = 1$$

We have also seen that

$$\frac{d\xi}{dt} = \frac{1}{v_i}\frac{dn_i}{dt} \tag{5}$$

The rate of a reaction is often determined experimentally by measuring some property of the system which depends on the number of moles of reactants and/or products. If the reaction takes place at constant volume, the number of moles of reactant (or product) is proportional to the concentration. In this case,

$$\text{rate of reaction (in moles per L per unit time)} = \frac{d\{\xi/V\}}{dt} = \frac{1}{v_i}\frac{d(i)}{dt} \tag{6}$$

where (i) is the concentration of i in moles/L.

Critical Thinking Questions

8. Show that equation (5) is consistent with your answer to CTQ 7.

9. Show that the rate of reaction, the right-hand-side of equation (5), is always positive.

Information

Consider the following reaction:

$$NH_4^+(aq) + NO_2^-(aq) \rightleftharpoons N_2(g) + 2\,H_2O(\ell) \tag{7}$$

Figure 1 presents a plot of the nitrite ion concentration versus time for this reaction. This plot clearly illustrates that the nitrite ion concentration changes very quickly at the beginning of the reaction and more slowly in the latter stages of the reaction. That is, the numerical value of $\frac{\Delta(\text{nitrite})}{\Delta\text{time}}$ is not constant.

A better measure of the rate of a reaction is the *instantaneous rate of reaction*, generally written as

$$\text{rate} = \frac{1}{v_i}\frac{d(i)}{dt} \tag{8}$$

Figure 1. Nitrite concentration versus time for the reaction of ammonium ion with nitrite ion

$$(NO_2^-)_0 = 0.00500 \text{ M} \qquad (NH_4^+)_0 = 0.100 \text{ M}$$

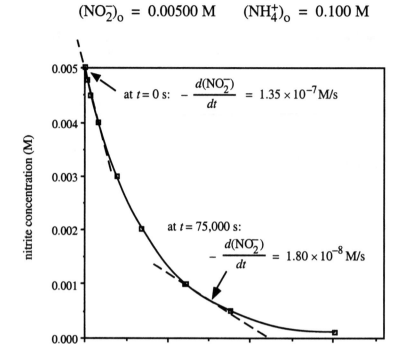

R.S. Moog and J.J. Farrell, *Chemistry: A Guided Inquiry*, 2nd Ed., John Wiley & Sons, Inc., 2nd Ed., 2002, CA 57. This material is used by permission of John Wiley & Sons, Inc.

Critical Thinking Questions

10. What is the rate of reaction at $t = 0$ s?

11. What is the rate of reaction at $t = 75,000$ s?

12. How does the rate of reaction change as (NO_2^-) decreases?

13. Estimate the value of the rate of reaction at $t = 175,000$ s. Explain your reasoning.

Model 2: The Effect of Concentration on Reaction Rate.

$$NH_4^+(aq) + NO_2^-(aq) \rightleftarrows N_2(g) + 2 H_2O(l)$$

Table 2. Initial reaction rates for several experiments at 25°C.

Experiment	Initial Concentration of NH_4^+ (M)	Initial Concentration of NO_2^- (M)	Initial Rate of Reaction (M /s)
1	0.100	0.0050	1.35×10^{-7}
2	0.100	0.010	2.70×10^{-7}
3	0.200	0.010	5.40×10^{-7}

Information

Often the rate of reaction (at constant temperature) is found to be proportional to the concentration of a reactant raised to some power (usually an integer such as 0, 1, 2, ...). For example, if

$$rate = k\,(R)^x \qquad then,$$

$$\frac{\text{initial rate}_2}{\text{initial rate}_1} = \frac{k\,(R)_2^x}{k\,(R)_1^x} = \left(\frac{(R)_2}{(R)_1}\right)^x$$

where k is called the **rate constant** and $k = f(T)$
initial rate$_j$ = the initial rate of experiment j
$(R)_j$ = the initial concentration of the reactant R for experiment j

The **rate law** for a particular reaction is an empirically determined expression describing how the rate of reactions depends upon the concentrations of reactants and products (and possibly catalysts). Although this is not always the case, often the rate law for a reaction such as equation (2) has the form

$$rate = k(A)^\alpha(B)^\beta(C)^\gamma(D)^\delta \qquad\qquad (9)$$

where k is a constant (at a given temperature) known as the **rate constant**, and the exponents are the **order** of the reaction with respect to each of the components. For example, if the exponent for component i is 1, then the reaction is said to be first-order in i.

Table 3. Experimental rate laws for several chemical reactions

Reaction	Experimental Rate Law
$CH_3Br(aq) + OH^-(aq) \rightleftharpoons CH_3OH(aq) + Br^-(aq)$	rate $= k \, (CH_3Br)$
$2O_3(g) \rightleftharpoons 3O_2(g)$	rate $= k \, (O_3)^2 \, (O_2)^{-1}$
$2\,HI(g) \rightleftharpoons H_2(g) + I_2(g)$	rate $= k \, (HI)^2$
$NH_4^+(aq) + NO_2^-(aq) \rightleftharpoons N_2(g) + 2\,H_2O(l)$	rate $= k \, (NH_4^+) \, (NO_2^-)$
$BrO_3^-(aq) + 5Br^-(aq) + 6H^+(aq) \rightleftharpoons 3Br_2(aq) + 3H_2O$	rate $= k \, (BrO_3^-)(Br^-) \, (H^+)^2$
$CH_3CHO(g) \rightleftharpoons CH_4(g) + CO(g)$	rate $= k \, (CH_3CHO)^{3/2}$

Critical Thinking Questions

14. Based on the data in Table 3, is the order of a reaction with respect to a particular species always equal to its stoichiometric coefficient in the balanced chemical equation?

15. Based on the data in Table 3, which of the following **must** be true about the order of a reaction with respect to a component:

    a)   equal to the stoichiometric coefficient

    b)   an integer

    c)   non-zero

    d)   positive

    e)   none of the above

Exercises

4. Assuming that the rate law is of the form given in equation (9), determine the rate law for the reaction in Model 2 using the data in Table 2.

5. Estimate the order with respect to nitrite for reaction given in equation (7) from the data given in Figure 1. Explain why this graph can be used to estimate the order of the reaction with respect to nitrite, but that it provides essentially no information about the order of the reaction with respect to ammonium ion.

6. The following data were collected for the reaction:

$$2\,NO(g) + O_2(g) = 2\,NO_2(g)$$

Initial NO Concentration (mol/L)	Initial O_2 Concentration (mol/L)	Initial Rate of reaction (mol/Ls)
5.38×10^{-3}	5.38×10^{-3}	9.55×10^{-6}
8.07×10^{-3}	5.38×10^{-3}	2.15×10^{-5}
13.45×10^{-3}	5.38×10^{-3}	5.97×10^{-5}
8.07×10^{-3}	6.99×10^{-3}	2.80×10^{-5}
8.07×10^{-3}	9.69×10^{-3}	3.88×10^{-5}

a) What is the rate law for the reaction?

b) What is the experimental rate constant for the reaction?

(c) Predict the initial rate of reaction when the initial concentration of NO is 1.50×10^{-2} mol/L and the initial concentration of O_2 is 2.00×10^{-3} mol/L.

Integrated Rate Laws

(How Does Concentration Change with Time?)

Focus Question: **How does the half-life of a second order reaction with a single reactant change as the reaction proceeds?**

a) **half-life increases**

b) **half-life decreases**

c) **half-life does not change**

d) **can't tell without balanced equation**

Model 1: A First Order Reaction.

Consider a balanced chemical equation of the form:

$$A + \text{other reactants} \rightarrow \text{products} \tag{1}$$

which is experimentally found to have a rate law given by

$$\text{rate} = k(A) \tag{2}$$

Then, the concentration of A varies in time according to

$$(A) = (A)_o e^{-kt} \tag{3}$$

where $(A)_o$ is the concentration of A at time $t = 0$, and (A) is the concentration of A at time t.

Critical Thinking Questions

1. Obtain equation (3) by following these steps:

    a)   Write the equation for the instantaneous rate of reaction, in terms of (A). [Hint, see equation (8) in CA K1.] Combine this equation with equation (2) above, the first-order rate law. Rearrange to obtain a new equation relating (A) to time, t. (That is, obtain an equation which has a term involving (A) and d(A) on one side and a term involving time, dt, on the other side.)

    b)   Both sides of the equation from part a) must be integrated in order to obtain equation (3). To do this, the limits of integration must be determined. The limits for time are $t = 0$ and $t = t$. When time $= t$, (A) = (A). When time $= 0$, what is the symbol for the concentration of A?

    c)   Integrate both sides of the equation using the limits from part b) to obtain equation (3).

2. Explain why a plot of $\ln \dfrac{(A)}{(A)_0}$ vs. t results in a straight line for the first order reaction, described in Model 1.

3.   How can the rate constant, k, be obtained from a plot of $\ln \dfrac{(A)}{(A)_o}$ vs. t for a first order reaction such as described in equation (1)?

Model 2: A Second Order Reaction.

Consider a balanced chemical equation of the form:

$$B + \text{other reactants} \rightarrow \text{products} \tag{4}$$

which is experimentally found to have a rate law given by

$$\text{rate} = k(B)^2 \tag{5}$$

Then, the concentration of B varies in time according to

$$\frac{1}{(B)} = \frac{1}{(B)_o} + kt \tag{6}$$

where $(B)_o$ is the concentration of B at time $t = 0$, and (B) is the concentration of B at time t.

Critical Thinking Questions

4.   What function of (B) may be plotted vs. what function of t to obtain a straight line plot for a second order reaction, according to equation (6)?

5.   Describe how the rate constant for a second order reaction may be obtained from the plot described in CTQ 4 above.

Exercises

1.   Derive equation (6) using equation (5) and equation (8) from CA K1.

2.   (a) What are the units of the rate constants of first- and second-order reactions if the concentrations are expressed in moles per liter and the time in seconds? (b) If the rate of a reaction follows the rate equation rate $= (A)^{1/3} (B)$, what are the units of k?

Information

The half-life, $t_{1/2}$, of a reaction with a single reactant is the time that it takes for the concentration of the reactant to reach one-half of its original value.

Critical Thinking Questions

6. In Model 1, what is the concentration of A at $t = 0$?

7. In Model 1, what is the concentration of A at $t = t_{1/2}$?

8. Show that $t_{1/2} = \dfrac{\ln 2}{k}$ for a first order reaction.

9. Derive an expression for $t_{1/2}$ for a second order reaction which indicates that the half-life depends on the initial concentration of reactant.

Exercises

3. The rate law for a reaction is known to involve only the reactant A, and is suspected to be either first-order or second-order. Describe, using grammatically correct English sentences, how the order of the reaction can be determined by measuring how long it takes for the concentration of A to reach 50% and 25% of its original value.

4. Which graph (I, II, III, IV, V) best describes the following reaction if the reaction is first order in N_2O_4?

$$N_2O_4(g) \rightleftharpoons 2\,NO_2(g)$$

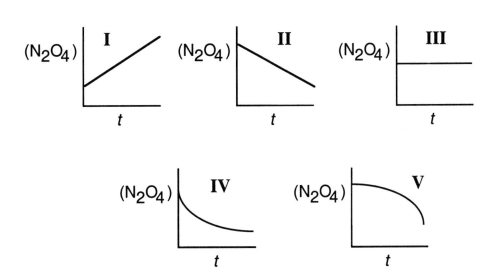

ChemActivity K3

The Method of Isolation for Determining Rate Laws
(How Can a Rate Law Be Obtained Experimentally?)

Focus Question: Consider the reaction

$$O_3(g) + NO(g) \rightarrow NO_2(g) + O_2(g)$$

which has been experimentally determined to have the rate law:

$$\text{rate} = k \, (O_3) \, (NO) \quad .$$

Under what experimental conditions will this reaction appear to be only first order in NO?

a) large excess of O_3

b) large excess of NO

c) $(NO)_0 = (O_3)_0$

d) $(O_3)_0 = 0$

Model 1: The Reaction of O_3 and NO.

One of the reactions which may be involved in the destruction of the ozone layer in the atmosphere is

$$O_3(g) + NO(g) \rightarrow NO_2(g) + O_2(g) \tag{1}$$

which has been experimentally determined to have the rate law:

$$\text{rate} = k \, (O_3) \, (NO) \tag{2}$$

In a particular experiment in the laboratory, the initial conditions in a closed 1.00-L vessel were:

$$(O_3)_o = 0.100 \text{ M} \qquad (NO)_o = 1.00 \times 10^{-4} \text{ M}$$

Critical Thinking Questions

1. For the particular experiment described in Model 1 above, explain why $(O_3) \approx (O_3)_0$ for all time.

2. For the particular experiment described in Model 1 above, if we write the rate as follows:
   $$\text{rate} = k' \, (NO)$$

   what is the expression for k'?

3. What type of simple rate law does the relationship in CTQ 2 resemble?

4. Propose experimental conditions which would result in the observed rate law for the reaction to be
   $$\text{rate} = k'' \, (O_3)$$

Information

We have seen that under certain experimental conditions, the mixed second-order rate law for the reaction in Model 1 can have the functional form of a simple first-order rate law:

$$\text{when } (O_3)_o \gg (NO)_o \qquad \text{rate} = k' \, (NO) \qquad (3)$$

$$\text{when } (NO)_o \gg (O_3)_o \qquad \text{rate} = k'' \, (O_3) \qquad (4)$$

The rate laws presented in equations (3) and (4) are called **pseudo first-order** rate laws. Determining the order of a reaction with respect to one component by performing the reaction under conditions in which all other components are present in large excess is called the **method of isolation**.

Critical Thinking Questions

5. k' in equation (3) is referred to as a **pseudo-order rate constant**, or **apparent rate constant**. Using grammatically correct sentences define pseudo-order rate constant.

6. The rate law for the reaction

$$cC + dD \rightarrow \text{products}$$

has an expected rate law of the form: $\text{rate} = k\,(C)^{\alpha}(D)^{\beta}$

The orders α and β are unknown.

a) Under what conditions will the rate law for this reaction be a pseudo β-order law in (D)?

b) Assume that β is equal to either 1 or 2. Using grammatically correct English sentences, explain how the value of β could be obtained from a series of experiments using the method of isolation. Include a discussion of both the experiments which would be done and how the data would be analyzed.

Exercises

1. The decomposition of hydrogen peroxide is catalyzed by iodide ion.

$$2\,H_2O_2 \rightarrow 2\,H_2O + O_2$$

   Iodide is not consumed by the reaction. The following data were obtained for the decomposition of hydrogen peroxide in 0.02 M KI at 25°C.

Time (min)	Volume of O_2 Evolved (mL)
0.00	0.00
5.00	7.50
10.00	14.00
25.00	28.80
45.00	41.20
65.00	48.30
∞	57.90

   Determine the pseudo order with respect to H_2O_2 and find the value of the apparent rate constant at this temperature. Assume that the reverse reaction can be neglected.

2. In acidic solution, ethanol is oxidized by $HCrO_4^-$ to produce acetic acid and Cr^{3+}. Under the conditions when both H^+ and ethanol are present in large excess, the rate of change of the chromium species may be written

$$-\frac{d(HCrO_4^-)}{dt} = \frac{d(Cr^{3+})}{dt} = k'\,(HCrO_4^-)^\alpha$$

   where k' is a pseudo rate constant and α is the order of the reaction with respect to $HCrO_4^-$.

   A typical physical chemistry experiment is to determine the value of α by comparing the experimentally determined time dependence of absorbance (for either of the species) to the expected dependence from the integrated rate law.

   Assuming that when the reaction begins $(HCrO_4^-) = C_o$ and $(Cr^{3+}) = 0$, determine the expression for (Cr^{3+}) as a function of time if the reaction is first order with respect to $HCrO_4^-$.

Reaction Mechanism

(What is a Mechanism?)

Focus Question: **Which of the following are bimolecular elementary processes?**

a) $CH_3Cl \rightarrow CH_3 + Cl$

b) $CH_3 + Cl \rightarrow CH_3Cl$

c) $Br_2 \rightarrow Br + Br$

d) $2\,NO + H_2 \rightarrow N_2O + H_2O$

e) $2\,H \rightarrow H_2$

Information

A **mechanism** is a sequence of chemical steps which produces some overall chemical transformation. Each of these steps is referred to as an **elementary step**. An elementary step usually involves the collision of two chemical species, or the dissociation of a single species into more than one component.

It is possible to propose more than one series of elementary steps which results in some overall chemical transformation. A proposed mechanism is a *possibly* correct description of the actual chemical steps if it predicts the experimentally observed rate law.

Model 1: A Proposed Mechanism.

Overall chemical transformation:

$$H_2(g) + 2\,NO(g) \rightarrow N_2O(g) + H_2O(g) \qquad (1)$$

Experimentally observed rate law:

$$\text{rate} = k\,(NO)^2\,(H_2) \qquad (2)$$

Proposed mechanism:

$$NO + NO \overset{k_1}{\underset{k_{-1}}{\rightleftarrows}} N_2O_2 \qquad\qquad \text{Step 1}$$

$$N_2O_2 + H_2 \overset{k_2}{\rightarrow} N_2O + H_2O \qquad \text{Step 2}$$

where
  k_1 is the rate constant for the bimolecular (forward) step 1,
  k_{-1} is the rate constant for the unimolecular (reverse) step 1,
  k_2 is the rate constant for the bimolecular step 2.[1]

Critical Thinking Questions

1. Why is the forward step 1 referred to as a bimolecular step, whereas reverse step 1 is referred to as a unimolecular step?

2. Write an expression (or expressions) which shows how the overall rate of reaction is related to the rate of production or consumption of each of the reactants and products in equation (1). For example, for H_2,

$$\text{rate of reaction} = -\frac{d(H_2)}{dt}$$

---

[1]In principle every step is reversible (as in step 1). However, if the reverse step has an extraordinarily small rate constant it is not explicity given in the mechanism.

Information

Unlike a balanced chemical equation, each elementary step in a proposed mechanism describes a specific process which is occurring on a molecular level. Thus, the rate law for any elementary step is completely determined by the reactants in the step. Furthermore, for a bimolecular step the rate is proportional to the number of collisions (only a certain percentage of collisions are successful) and it can be shown that the number of collisions is proportional to the product of the concentrations. Thus, for the forward step 1 in Model 1:

$$\text{rate of forward step 1} = k_1 \, (NO) \, (NO) \; = \; \left\{ \frac{d(N_2O_2)}{dt} \right\}_{\text{step 1, forward}} \tag{3}$$

For a unimolecular elementary process the rate depends on the number of molecules. [There will be a certain probability that any one molecule will have sufficient energy in a particular vibrational mode to undergo the decomposition.] Thus, the rate simply depends on the concentration of that molecule. For the reverse step in Model 1,

$$\text{rate of reverse step 1} \; = k_{-1} \, (N_2O_2) \; = \; \left\{ -\frac{d(N_2O_2)}{dt} \right\}_{\text{step 1, reverse}} \tag{4}$$

Critical Thinking Questions

3. Write an expression for the rate of step 2 in Model 1, and show how it is related to the change of (N_2O_2) with time as in equations (3) and (4).

4. Write expressions for each of the following:

    a)   relationship of rate of forward step 1 to $\dfrac{d(NO)}{dt}$

    b)   relationship of rate of reverse step 1 to $\dfrac{d(NO)}{dt}$

c)   relationship of rate of step 2 to $\dfrac{d(N_2O)}{dt}$, $\dfrac{d(H_2)}{dt}$, $\dfrac{d(H_2O)}{dt}$

5.   From equation (1) we know that

$$\text{rate of reaction} = \frac{d(H_2O)}{dt}$$

Explain why this implies, according to the mechanism in Model 1, that the rate of reaction = $k_2 \, (N_2O_2) \, (H_2)$.

6. Is N_2O_2 a reactant or a product in the overall chemical transformation described in equation (1)?

7. Based on the expression for the rate of reaction in CTQ 5, is the proposed mechanism in Model (1) a possibly correct mechanism? Why or why not?

Information

For the mechanism described in Model 1, N_2O_2 is neither a reactant nor a product. It is an **intermediate** — it is produced and consumed within the mechanism, and does not occur in the stoichiometric equation.

Model 2: The Steady State Approximation.

It is difficult to tell if the proposed mechanism in Model (1) is a possibly correct mechanism based on the rate of reaction given in CTQ 5 because the expression there includes (N_2O_2), an intermediate species. In general, rate laws derived from mechanisms should have only have terms which include the reactants and/or products of the overall reaction, because these are the species which are most readily measured experimentally. A widely used method for eliminating terms involving reaction intermediates from these expressions is the **steady state approximation**. This approximation states that the concentration of a reaction intermediate is essentially unchanged during the course of a reaction. For the proposed mechanism in Model 1, this would be mathematically represented as :

$$\left\{ \frac{d(N_2O_2)}{dt} \right\}_{total} = 0 \tag{5}$$

Another way of stating this is that the overall rate of production of N_2O_2 is equal to the overall rate of consumption of N_2O_2.

Critical Thinking Questions

8. In Model 1, what is the expression for the rate of production of N_2O_2 in forward step 1 (in terms of k_1 and (NO))?

9. In Model 1, what is the analogous expression for the rate of consumption of N_2O_2 in reverse step 1?

10. In Model 1, what is the expression for the rate of consumption of N_2O_2 in step 2?

11. What is the total rate of consumption of N_2O_2 ?

12. Provide an expression in which the rate of production of N_2O_2 is set equal to the rate of consumption of N_2O_2.

13. Use your result from CTQ 12, and the expression for the rate of reaction from CTQ 5 to show that the predicted rate of reaction from the mechanism in Model 1 is

$$\text{rate} = \frac{k_2 k_1 \, (NO)^2 \, (H_2)}{k_{-1} + k_2 \, (H_2)}$$

14. In order for the proposed mechanism to be consistent with the experimentally determined rate law,

   a) what must be true about the relative values of the two terms in the denominator of the expression in CTQ 13?

   b) What would this imply about the relative rates of the elementary steps in the mechanism? That is, in order for the mechanism to be possibly correct, which step (or steps) in the mechanism must be much faster (or much slower) than what other step (or steps)?

   c) Based on your answer to part b) above, which (if any) of the steps in the proposed mechanism would be considered a "rate-determining step"?

15. An experimentally determined rate law is always based on a finite number of experiments, covering a variety (but not all possible) experimental conditions.

    In order to test whether the proposed mechanism in Model 1 is truly "correct", what other experiments could be done to further support (or refute) it?

    Hint: Consider under what conditions the rate law would be predicted to change, and in what way.

Exercises

1. The following is elementary step 3 in a mechanism:

   $$A + B \xrightarrow{k_3} C$$

   a) What is the rate of the step? b) How is $-d(A)/dt$ for this step related to the rate of this step? c) How is $d(C)/dt$ for this step related to the rate of this step?

2. The following is elementary step 2 in a mechanism:

   $$A + A \xrightarrow{k_2} C$$

   a) What is the rate of the step? b) How is $-d(A)/dt$ for this step related to the rate of this step? c) How is $d(C)/dt$ for this step related to the rate of this step?

3. The thermal decomposition of ethane leads to a mixture of H_2, CH_4, and C_2H_4, with small amounts of higher hydrocarbons. One proposed mechanism consists of the steps:

   $$C_2H_6 \xrightarrow{k_1} CH_3 + CH_3$$

   $$CH_3 + C_2H_6 \xrightarrow{k_2} CH_4 + C_2H_5$$

   $$C_2H_5 \xrightarrow{k_3} C_2H_4 + H$$

   $$H + C_2H_6 \xrightarrow{k_4} H_2 + C_2H_5$$

$$C_2H_5 + C_2H_5 \xrightarrow{k_5} C_4H_{10}$$

$$C_2H_5 + C_2H_5 \xrightarrow{k_6} C_2H_4 + C_2H_6$$

Apply the steady state approximation to determine expressions for the rates of production of methane, CH_4, and of dihydrogen, H_2, and show that CH_4 production is first-order in ethane, and that H_2 production is one half-order in ethane.

4. Over a wide range of experimental conditions, the gas-phase reaction

$$2\,NO + O_2 \rightarrow 2\,NO_2$$

has been observed to be second order in NO and first order in O_2. Two proposed mechanisms for this conversion are shown below:

MECHANISM A:

$$NO + NO \underset{k_{-A1}}{\overset{k_{A1}}{\rightleftarrows}} N_2O_2$$

$$N_2O_2 + O_2 \xrightarrow{k_{A2}} NO_2 + NO_2$$

MECHANISM B:

$$NO + O_2 \underset{k_{-B1}}{\overset{k_{B1}}{\rightleftarrows}} NO_3$$

$$NO_3 + NO \xrightarrow{k_{B2}} NO_2 + NO_2$$

(a) Use the steady state approximation to derive the predicted rate law for each of these mechanisms. Describe the restrictions that must be placed on the relative rates of the steps in each mechanism in order for each to be an acceptable possible mechanism for the reaction.

(b) Describe an experiment (or set of experiments) which would enable the differentiation of the two mechanisms proposed.

5. The thermal decomposition of acetaldehyde (CH_3CHO) is thought to follow the Rice-Herzfeld mechanism shown below:

$$CH_3CHO \xrightarrow{k_1} CH_3 + CHO$$

$$CH_3 + CH_3CHO \xrightarrow{k_2} CH_4 + CH_2CHO$$

$$CH_2CHO \xrightarrow{k_3} CO + CH_3$$

$$CH_3 + CH_3 \xrightarrow{k_4} C_2H_6$$

Apply the steady state approximation to determine expressions for the rate of production of methane, CH_4, and of ethane, C_2H_6.

6. Consider the following mechanism for the conversion of A to P:

$$A + A \underset{k_{-1}}{\overset{k_1}{\rightleftarrows}} B$$

$$B \xrightarrow{k_2} P$$

This mechanism suggests that the overall conversion is $2\,A \rightarrow P$ and predicts that the overall rate of reaction is second order in A.

a) Show that if the final step in the mechanism is **replaced** with the alternative step

$$B + A \xrightarrow{k_{2'}} P$$

(making the overall reaction $3\,A \rightarrow P$) that the obtained rate law is

$$\text{rate} = \frac{k_1\, k_{2'}\, (A)^3}{k_{-1} + k_{2'}\, (A)}$$

b) The conversion of A to P was observed to be second order in A under a certain set of experimental conditions. Discuss the implications of this result in terms of any restrictions that it places on the relative rates of the various steps in the mechanism proposed in part (a) above. In addition, describe an experiment (or set of experiments) which would enable the differentiation of the two mechanisms proposed.

K5

Activation Energy
(How Does a Rate Constant Depend on Temperature?)

Focus Question: Which of the following graphs best describes how the rate constant for a typical chemical reaction depends on temperature?

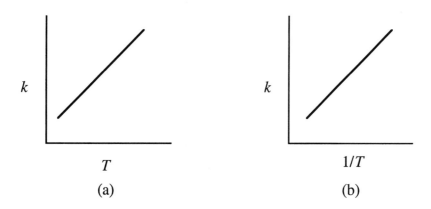

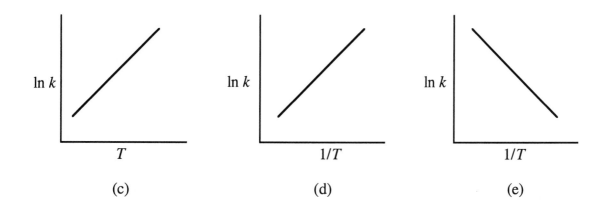

Information

The rate of a chemical reaction usually shows a very strong dependence on temperature. Although most reactions exhibit the same general form for the dependence of rate on temperature, not all do. Three examples of the dependence of rate on temperature are shown below in Model 1.

Model 1: Examples of Dependence of Rate on Temperature.

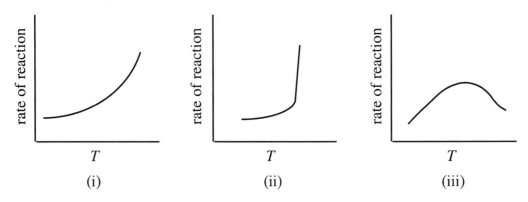

These three figures show examples of:
   a) a reaction that reaches an explosive stage at some temperature
   b) an enzyme catalyzed reaction in which the enzyme becomes deactivated at high temperature,
   c) a typical chemical reaction.

Critical Thinking Questions

1.   For the three types of reactions described in Model 1, which corresponds to which plot?

2.   Based on the appropriate plot in Model 1, use grammatically correct English sentences to describe how rate depends on temperature for a typical chemical reaction.

Information

We have seen that often the rate law for a reaction has the form

$$\text{rate} = k(A)^{\alpha}(B)^{\beta}(C)^{\gamma}(D)^{\delta}$$

where A, B, C, and D are reactant or products and the exponents are the order of the reaction with respect to each of the components. The concentration dependence of a rate law (the part involving the concentration and order of the species) is rarely temperature dependent; the rate constant of a reaction almost always is temperature dependent.

Model 2: Temperature Dependence of a Rate Constant.

The gas-phase decomposition of acetaldehyde, CH_3CHO, is observed to be second order over the temperature range 600 - 1000 K. Measured rate constants over this temperature range are reported in Table 1.

Table 1: Rate Constants for Decomposition of Acetaldehyde as a Function of Temperature.

T (K)	k (Lmol^{-1}s^{-1})
600	0.000054
700	0.011
800	0.535
900	11.5
1000	145.

Critical Thinking Questions

3. Based on the data in Table 1, and without using a calculator, determine which (if any) of the following functional forms is likely to be a good description of the dependence of k on T, where A and B are positive-valued constants:

    a) $k = A \times T$

    b) $k = A / T$

    c) $k = A\, e^{BT}$

    d) $k = A\, e^{B/T}$

    e) $k = A\, e^{-B/T}$

4. For <u>each of the two likely functional forms</u> in CTQ 3:

   a)   What function of k should be plotted vs. what function of T in order to get a straight line?

   b)   Use grammatically correct English sentences to describe how the value of A (and B, if appropriate) can be obtained from a fit to each line.

5. Make appropriate plots for the two likely functional forms from CTQ 3 using the data in Table 2 and the graph paper on the next page. Based on these plots, what is the functional form for the relationship between k and T for the decomposition of acetaldehyde?

Table 2: Relevant Data for Acetaldehyde Decomposition

$T(K)$	k (mole L^{-1} s^{-1})	$1/T$ (K^{-1})	$\ln k$
600	0.000054	0.00167	-9.83
700	0.011	0.00143	-4.51
800	0.535	0.00125	-0.625
900	11.5	0.00111	2.442
1000	145.	0.00100	4.977

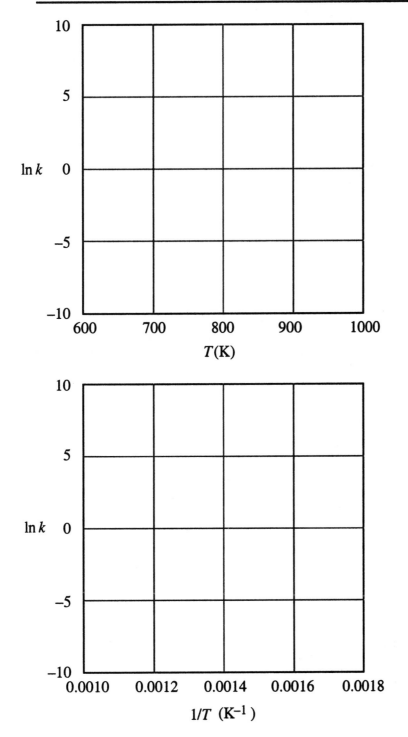

Information

Most chemical reaction rates exhibit a temperature dependence which closely follows the functional form found for the decomposition of acetaldehyde above. This relationship is generally expressed as the Arrhenius equation

$$k = A\, e^{-E_a/RT} \tag{1}$$

or

$$\ln k = \ln A - \frac{E_a}{RT} \tag{2}$$

where R is the gas constant, A is the pre-exponential factor (or frequency factor), and E_a is the activation energy. A and E_a are often referred to as the Arrhenius parameters.

Critical Thinking Questions

6. Based on equation (1), what must be the units for A for the rate constant of a first-order reaction? A second-order reaction?

7. Many chemical reactons have E_a ranging from 30 kJ/mole to 150 kJ/mole. For which type of reaction will a given change in temperature (near room temperature) result in a larger change in the rate of reaction: a reaction with E_a toward the low end of this range or with E_a toward the high end of this range? Explain your reasoning.

Exercises

1. Determine the values for E_a and A for the decomposition of acetaldehyde from the data in Table 2.

2. The number of chirps per minute of a snowy tree cricket (*Oecanthus fultoni*) at several temperatures is given below:

T (°C)	Chirps per minute
25.0	178
20.3	126
17.3	100

   Find the activation energy for the chirping rate and the expected chirping rate at 14.0 °C. Compare these results with the well-known rule that the Fahrenheit temperature may be determined by taking the number of chirps in 15 seconds and adding 40. [°F = 1.8 (°C) + 32]

3. If a rate constant were found to triple when the temperature is increased from 25°C to 35°C, what is the value of E_a?

Activation Energy(II)

Focus Question: If the thermodynamic parameters for a chemical reaction are favorable, will the reaction occur?

Model 1: The Reaction Coordinate Diagram for the Reaction $N_2O(g) + NO(g) \rightleftharpoons N_2(g) + NO_2(g)$.

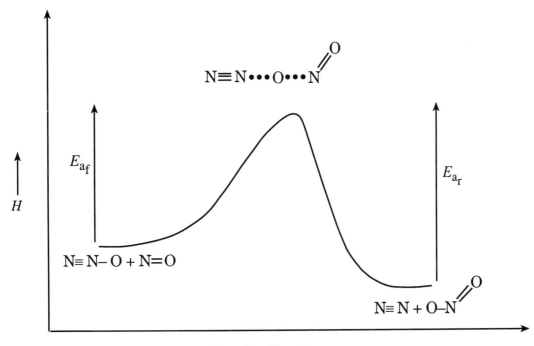

Reaction Coordinate

According to the Arrhenius formulation, before reactants can be converted into products the collision energy of the system must overcome the activation energy. The vertical axis gives the enthalpies (activation energies are usually taken to be the same as enthalpies) and the horizontal axis represents the sequence of infinitesimally small changes that must occur to convert reactants into products. Molecules must collide with sufficient energy and in the proper orientation relative to one another to cause a reaction to occur. The collision complex, called the transition state or the activated complex, is shown above at the top of the activation energy barrier.

Critical Thinking Questions

1. Calculate the enthalpy change for the chemical reaction of Model 1.

   $\Delta_f H^{\circ}_{N_2O}(g) = 82.05$ kJ/mol $\Delta_f H^{\circ}_{NO_2}(g) = 33.2$ kJ/mol

   $\Delta_f H^{\circ}_{NO}(g) = 91.3$ kJ/mol

2. According to Model 1, is the activation energy for the reverse reaction the same, greater than, or less than the forward activation energy?

3. Note that $E_{a_f} = H^{\ddagger} - H^{\circ}_{reac}$ where $H^{\ddagger}$ represents the enthalpy of the activated complex and show that:

   $$\Delta_r H = E_{a_f} - E_{a_r}$$

   Hint: Recall $\Delta_r H = H_{prod} - H_{reac}$

4. The experimental activation energy in the forward direction in Model 1 is 209 kJ/mol. What is the value of the activation energy in the reverse direction?

Model 2: A Proposed Mechanism for 2NO(g) + O₂(g) ⇌ 2NO₂(g).

The reaction mechanism has been proposed to be

$$NO + NO \xrightarrow{k_1} N_2O_2$$
$$N_2O_2 \xrightarrow{k_{-1}} 2NO$$
$$N_2O_2 + O_2 \xrightarrow{k_2} 2NO_2$$

and the experimental rate law is

$$\frac{d(NO_2)}{dt} = k(NO)^2(O_2)$$

where k is the overall rate constant. (See CA K4, Exercise 4.)

The rate law predicted from the mechanism is

$$\text{rate} = \frac{1}{2}\frac{d(NO_2)}{dt} = k_2\frac{k_1}{k_{-1}}(NO)^2(O_2)$$

Critical Thinking Questions

5. Identify the overall rate constant, k, with the rate constants for each step of the mechanism.

6. For each individual rate constant substitute the corresponding Arrhenius expression and show that

$$k = \frac{A_2 A_1}{A_{-1}} e^{-(E_{a_2} + E_{a_1} - E_{a_{-1}})/RT}$$

7. If $E_{a_1} = E_{a_2} = 82$ kJ/mol and $E_{a_{-1}} = 205$ kJ/mol, calculate the overall activation energy for the reaction. What is unusual about the overall activation energy?

Model 3: A $\rightleftharpoons$ B.

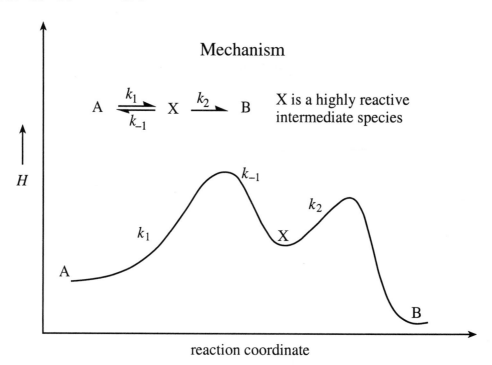

Mechanism

$$A \underset{k_{-1}}{\overset{k_1}{\rightleftharpoons}} X \xrightarrow{k_2} B$$

X is a highly reactive intermediate species

Critical Thinking Questions

8. For the model reaction, what is $\dfrac{d(B)}{dt}$?

9. Apply the steady state assumption to find (X).

10. What is the rate law for this mechanism?

11.   a)   According to the model which is larger E_{a_1} or E_{a_2}?

       b)   Based on your answer to part a), which is a more reasonable approximation: $k_2 \gg k_{-1}$ or $k_{-1} \gg k_2$?

12.   Use the more reasonable approximation from CTQ 11 to rewrite the rate law in a simpler form.

13.   What step, if any, determines the rate of this reaction?

14.   Draw a reaction coordinate diagram for the model reaction if $E_{a_2} \gg E_{a_1}$.

15.   Begin with the rate of reaction as determined in CTQ 10, consider the possible approximations from CTQ 11, and rewrite the rate law based on the reaction coordinate diagram of CTQ 14.

16. If the steps A $\underset{k_{-1}}{\overset{k_1}{\rightleftharpoons}}$ X are fast equilibrium steps, what thermodynamic quantity is given by the ratio k_1/k_{-1}?

17. Rewrite the rate law from CTQ 15 based on your answer to CTQ 16.

18. According to the coordinate diagram of CTQ 14, is X reconverted to A faster than X is converted to product? What is the rate limiting step based on this observation and on your answer to CTQ 17?

19. If it is true that $k_{-1} \gg k_2$ does the relationship of k_2 to k_1 matter?

20. Is the step with the largest activation energy always the rate limiting step?

Exercises

1. For the thermal decomposition of $C_2H_6(g)$ the chain reaction mechanism yields the following rate law:

$$\frac{-d(C_2H_6)}{dt} = \left(\frac{k_1 k_3 k_4}{k_5}\right)^{1/2} (C_2H_6)$$

If the activation energies are

E_{a_1}	351 kJ/mol
E_{a_3}	167
E_{a_4}	29
E_{a_5}	0

what is the overall activation energy for the reaction?

2. For $A(g) \underset{k_{-1}}{\overset{k_1}{\rightleftharpoons}} C(g)$

$k_1 = 0.10 \text{ s}^{-1}$ (298 K) $k_{-1} = 2.0 \times 10^{-3} \text{s}^{-1}$

$k_1 = 0.25 \text{ s}^{-1}$ (313 K) $k_{-1} = 4.0 \times 10^{-3} \text{s}^{-1}$

Calculate the equilibrium constant at 298 K, E_{a_f}, E_{a_r}, and $\Delta_r H$.

3. Rationalize the activation energy found for CTQ 7.

Enzyme Kinetics
(What is a Catalyst?)

Focus Question: **A typical catalytic reaction involving a biological enzyme converts a substrate into a product. A plot of rate vs. substrate concentration is shown below.**

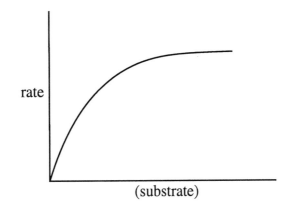

Provide an explanation for the shape of this curve.

Information

A **catalyst** is substance which increases the rate of a reaction but is neither produced nor consumed in the overall chemical process.

Model 1: Michaelis-Menten Catalytic Mechanism.

Many biological transformations are catalyzed by enzymes. The simplest catalytic mechanism for enzymes is the Michaelis-Menten mechanism.

Overall chemical transformation:

$$S \rightarrow P \tag{1}$$

$$\text{rate} = \frac{d(P)}{dt} \tag{2}$$

Proposed mechanism:

$$S + E \underset{k_{-1}}{\overset{k_1}{\rightleftarrows}} E{\cdot}S \tag{step 1}$$

$$E{\cdot}S \overset{k_2}{\rightarrow} P + E \tag{step 2}$$

where
k_1 is the rate constant for the forward process in step 1;
k_{-1} is the rate constant for the reverse process in step 1;
k_2 is the rate constant for the forward process in step 2;
S is the **substrate** (reactant)
P is the product
E is the **enzyme** (catalyst)
E${\cdot}$S is an enzyme-substrate complex which can
  - dissociate to revert back to the original substrate   or
  - undergo a transformation to produce product and the original enzyme

$$\text{rate} = k_2\,(E{\cdot}S) = \frac{k_2 k_1}{k_{-1} + k_2}\,(E)\,(S) \tag{3}$$

Critical Thinking Questions

1. Why is the rate of reaction in equation (3) equal to the rate of step 2 of the mechanism? That is, why is rate = k_2 (E•S) ?

2. What is the intermediate in the Michaelis-Menten mechanism?

3. Use the steady state approximation on (E•S) to show that

$$(E•S) = \frac{k_1 \, (S) \, (E)}{k_{-1} + k_2} \tag{4}$$

4. Show that the rate $= \dfrac{k_2 \, k_1}{k_{-1} + k_2}$ (E) (S)   (5)

Information

Note that in equation (3), the rate of the reaction is expressed in terms of either the concentration of an intermediate, (E•S), or in terms of the concentration of enzyme, (E). Neither of these quantities is readily measurable in typical laboratory situations. One reason for this is that it is very difficult to find an experimental technique which can clearly distinguish between free enzyme (E) and the enzyme-substrate complex (E•S). Typically, however, the total enzyme concentration, $(E)_{tot}$, is known from the preparation of the experimental sample.

Generally, the amount of substrate present is much greater than the amount of enzyme. For this reason, it is a reasonable approximation to assume that $(S) = (S)_o$ at all times, where $(S)_o$ is the initial concentration of S.

Critical Thinking Questions

5. Assuming that all of the enzyme originally present in an experimental sample is either free enzyme or involved in an enzyme-substrate complex, what is the mathematical relationship between $(E)_{tot}$, (E), and $(E \cdot S)$?

6. Explain why it is reasonable to assume that $(S) = (S)_o$ at all times.

7. Substitution of $(E) = (E)_{tot} - (E \cdot S)$ (from CTQ 5) into equation (4) yields

$$(E \cdot S) = \frac{k_1(S)(E)_{tot}}{k_{-1} + k_2 + k_1(S)}.$$ Show that the rate $= \frac{k_2 k_1(S)(E)_{tot}}{k_{-1} + k_2 + k_1(S)}.$

Information

For a given combination of enzyme and substrate, it is useful to define a Michaelis constant, K_M:

$$K_M = \frac{k_2 + k_{-1}}{k_1} \qquad (6)$$

Thus, this mechanism leads to the following expression for the rate of reaction, which is known as the **Michaelis-Menten equation**:

$$\text{rate} = \frac{k_2(S)(E)_{tot}}{K_M + (S)} \qquad (7)$$

Critical Thinking Questions

8. Based on your answer to CTQ 7 and equation (6), show that equation (7) is correct.

9. Consider the situation in which (S) $\ll K_M$. Under this situation, show that the rate is proportional to (S) for a given (that is, constant) $(E)_{tot}$.

10. Consider the situation in which (S) $\gg K_M$. Under this situation, show that the rate is independent of (S) for a given (constant) $(E)_{tot}$.

11. Based on your answers to CTQs 9 and 10, sketch a graph of rate vs. (S) for a given (constant) $(E)_{tot}$.

Information

For a given $(E)_{tot}$, the maximum rate of reaction is

$$\text{rate}_{max} = k_2 \, (E)_{tot} \tag{8}$$

The rate constant k_2 is referred to as the **catalytic constant** or **turnover number** for the enzyme-substrate pair. This can be thought of as the rate of reaction per mole of enzyme.

Many scientists have found it useful to graph $\dfrac{1}{\text{rate}}$ vs. $\dfrac{1}{(S)}$ for a variety of different (S) with a given $(E)_{tot}$. Such a plot is often referred to as a **Lineweaver-Burke plot**.

Critical Thinking Questions

12. Rearrange equation (7) to show that a Lineweaver-Burke plot should give a straight line.

13. What is the slope of a Lineweaver-Burke plot?

14. What is the intercept of a Lineweaver-Burke plot?

Exercises

1. Show that substitution of $(E) = (E)_{tot} - (E \cdot S)$ (from CTQ 5) into equation (4) yields

$$(E \cdot S) = \frac{k_1 \, (S) \, (E)_{tot}}{k_{-1} + k_2 + k_1(S)}.$$

2. The reaction of dissolved (in water) CO_2 to the bicarbonate ion is catalyzed by the enzyme carbonic anhydrase. The following results were obtained for a pH = 7 buffer solution at 1.5°C [*J. Am. Chem. Soc.*, **80**:5209 (1958)].

conc of CO_2 (mmol/L)	Initial rate (mmol/Ls)
0.76	0.0062
1.51	0.0116
3.78	0.0204
7.57	0.0287
15.1	0.0368

   Prepare a Lineweaver-Burke plot and determine the values of K_M and $k_2[E]_{tot}$.

3. Use equation (7) to show that K_M is the concentration of substrate which yields a rate of reaction equal to $\frac{1}{2}$ rate$_{max}$.

Model 2: Catalytic Action.

The decomposition of HI can proceed in the gas phase.

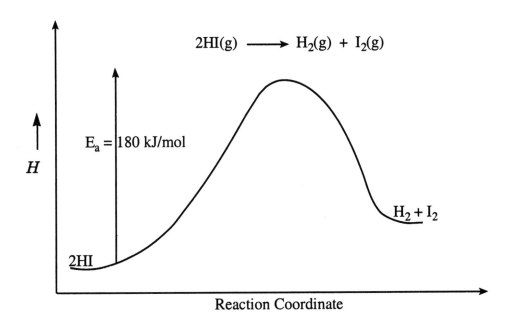

$$2HI(g) \longrightarrow H_2(g) + I_2(g)$$

$E_a = 180 \text{ kJ/mol}$

H

2HI

$H_2 + I_2$

Reaction Coordinate

An alternate mechanism in the presence of solid gold is that HI is chemisorbed on the gold surface, breaks apart to form H and I atoms, which then combine to produce H_2 and I_2. These in turn are desorbed from the surface.

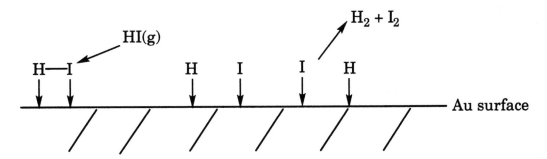

$H_2 + I_2$

HI(g)

H—I   H   I   I   H   Au surface

The activation energy for this process is 105 kJ/mol.

Critial Thinking Questions

15. On the reaction coordinate diagram above, draw the reaction coordinate diagram for the decomposition of HI on an Au surface. Label the diagram clearly.

16. Assuming that the Arrhenius pre-exponential factor A is about the same for both the gas phase and the Au surface catalyzed reactions, calculate the ratio of the rate constants for the catalyzed and uncatalyzed reaction at 300 K.

17. What effect does the presence of gold have on the rate of decomposition of HI?

18. Is the position of equilibrium affected by a catalyst?

19. Are the activation energies of both the forward and reverse reactions altered by the same amount? Explain.

Exercises

4. The catalase enzyme in blood or turnips catalyzes the decomposition of hydrogen peroxide. In the absence of a catalyst the activation energy for

$$2H_2O_2(aq) \rightleftharpoons 2H_2O(\ell) + O_2(g)$$

is 75.3 kJ/mol, while in the presence of catalase E_a is 8 kJ/mol. By what factor is the rate of reaction increased by catalase at 300 K?

5. The activation energies for a certain reaction are $E_{a_f} = 120$ kJ/mol, $E_{a_r} = 185$ kJ/mol. What would be the activation energy of the reverse reaction in the presence of a catalyst that decreased the E_{a_f} by 30 kJ/mol?

6. Suggest two pathways for the conversion of a biological substrate into a product. Explain how a catalyst may function to speed such a reaction.

7. For the enzyme catalyzed reaction:

$$E + S \underset{k_{-1}}{\overset{k_1}{\rightleftharpoons}} E{\bullet}S \underset{k_{-2}}{\overset{k_2}{\rightleftharpoons}} EP \underset{k_{-3}}{\overset{k_3}{\rightleftharpoons}} P + E$$

where $E{\bullet}S$ is highly reactive and $k_2 >> k_{-2}$, $k_3 >> k_{-3}$, roughly sketch the reaction coordinate diagram. P+E is more stable enthalpically than E+S and k_2 is rate limiting.

If $(E) \approx (E_o) - (E{\bullet}S)$ where E_o is the initial concentration of E, show that

$$(E{\bullet}S) = \frac{k_1(E_o)(S)}{k_{-1} + k_2 + k_1(S)} .$$

When will the maximum velocity of reaction occur; i.e., when is $V_{max} = k_2(E_o)$?

Collision Theory
of Gas Phase Reactions

Focus Question: A single molecule in the gas phase may experience 10^9 collisions per second. The total number of collisions between two different molecules is on the order of 10^{34} $m^{-3}s^{-1}$. Why do not all gas phase reactions proceed rapidly?

Model 1: Particles Must Collide in Order to React.

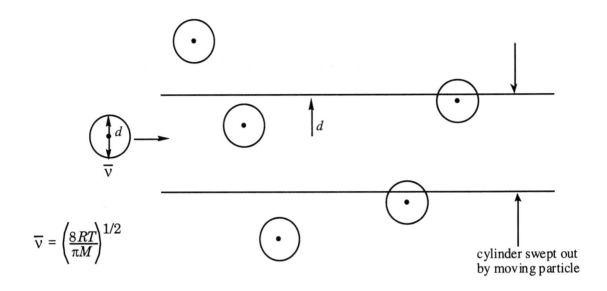

$$\overline{v} = \left(\frac{8RT}{\pi M}\right)^{1/2}$$

cylinder swept out
by moving particle

   A particle of diameter d moves through a collection of identical stationary particles. Any particle whose center falls within the cylinder swept out by the moving particle will be hit. Collision theory requires that particles must collide in order to react. The rate of the reaction then depends on the number of collisions that occur in unit time multiplied by the fraction of collisions that are effective in producing products.

Critical Thinking Questions

1. If the moving particle has speed equal to the average speed, $\overline{v}$ what is the expression for the volume of the cylinder swept out in one second as shown in the model?

2. The density of the stationary particles in the path of the moving particle is N/V, how many collisions per second will be experienced by the moving particle? Let Z_1 represent the total number of collisions per second.

Information

If all particles are moving, the expression developed in CTQ 2 is not correct. The relative velocity must be considered. It turns out that on average most collisions are at right angles.

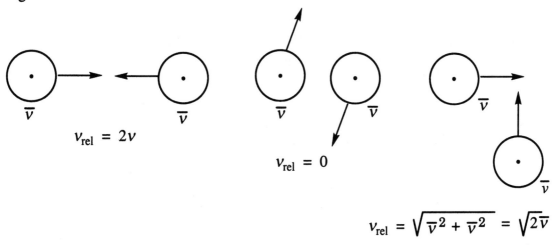

$$Z_1 = \sqrt{2}\ \bar{v}\pi d^2 \frac{N}{V} \tag{1}$$

Number of collisions per second experienced by a single particle.

The total number of collisions per unit volume per second experienced by all the molecules is given by

$$Z_{11} = \frac{1}{2} Z_1 \frac{N}{V} \tag{2}$$

where N is the total number of molecules in volume V.

Critical Thinking Questions

3. Explain how eqn (2) may be obtained from eqn (1). Pay particular attention to the factor of 2 that appears in eqn (2).

4. Use eqn (1) in eqn (2) to obtain an expression for Z_{11}.

5. If the particles are not identical the total number of collisions per second per unit volume between unlike particles is given the symbol Z_{12}. d_{12} is the average value of the radii of the two particles.

$$Z_{12} = \pi d_{12}^2 \left(\bar{v}_1^2 + \bar{v}_2^2 \right)^{1/2} \frac{N_1 N_2}{V^2}$$

Rationalize this equation and identify each term.

Model 2: The Rate of Reaction According to Collision Theory.

reaction rate = (collisions per unit time)(fraction of effective collisions)

Critical Thinking Questions

6. Use the above collision theory model to write an expression for the rate of a chemical reaction between unlike particles. Let C_1 and C_2 represent N_1/V and N_2/V. The fraction of effective collisions is not yet known.

7. Give the Arrhenius empirical relationship between rate constant and temperature. Give the rate law for a second order reaction. Use C_1 and C_2 to represent concentrations.

8. Identify the Arrhenius rate constant and the fraction of effective collisions by examination of CTQ 6 and 7.

Model 3: Effective and Ineffective Collisions.

$$N\equiv N-O(g) + N=O(g) \longrightarrow N\equiv N(g) + O-\overset{\overset{\displaystyle O}{\displaystyle \|}}{N}(g)$$

Sometimes when a N_2O molecule and a NO molecule collide, the collision does not lead to a N_2 molecule and a NO_2 molecule.

Critical Thinking Questions

9. Suggest a reason based on energy considerations why a collision between N_2O and NO might not produce N_2 and NO_2.

10. Suggest a reason why not all sufficiently energetic collisions between N_2O and NO will produce N_2 and NO_2.

Exercises

1. Calculate Z_1 and Z_{11} for O_2 gas at 0°C at 1 bar. $d = 3.6 \times 10^{-10}$ m.

2. The average distance a particle travels between collisions is called the mean free path. The mean free path depends on the number of collisions experienced by one particle in unit time and the average distance traveled in that unit time. Derive an expression to calculate the mean free path. What effect does increasing temperature have on λ? What effect does increased pressure have? Calculate the mean free path for O_2 at 0°C and 1bar.

3. At constant T and N a container is expanded to double its volume. By what factor does Z_{11} change? By what factor does Z_1 change?

Analysis of a
Gas Phase Chain Reaction

Focus Question: For a complex chain reaction, is it necessary that the rate constants for each step of the mechanism appear in the rate law derived from the mechanism?

Model 1: The Rate Law and Proposed Mechanism for the Gas Phase Reaction $H_2(g) + Br_2(g) \rightleftharpoons 2HBr(g)$.

The rate law for the reaction is experimentally found to be

$$\text{rate} = \frac{k(H_2)(Br_2)^{1/2}}{1 + m\frac{(HBr)}{(Br_2)}} \tag{1}$$

where k and m are constants at a fixed temperature.

The rate law may be derived from the following mechanism:

$$Br_2 + M \underset{k_5}{\overset{k_1}{\rightleftharpoons}} 2Br + M$$

$$Br + H_2 \underset{k_4}{\overset{k_2}{\rightleftharpoons}} HBr + H$$

$$H + Br_2 \overset{k_3}{\longrightarrow} HBr + Br$$

The symbol M may represent a wall, a molecule, or another atom. A third body collision partner must be present when two atoms combine to take away some of the energy of the combination. Thus in order for two bromine atoms to collide and remain bonded together a third body is needed.

Critical Thinking Questions

1. Using the above mechanism write an expression for the rate of reaction, $\frac{d(HBr)}{dt}$.

2. Apply the steady state assumption to the highly reactive radical H.

3. Find an expression for (H).

4. Apply the steady state assumption to the highly reactive radical species Br.

5. Find an expression for (Br). Hint: try adding CTQ's 2 and 4.

6. Show that

$$(H) = \frac{k_2(H_2)\left(\dfrac{k_1}{k_5}\right)^{1/2}(Br_2)^{1/2}}{k_3(Br_2) + k_4(HBr)}$$

7. Using the answers to CTQ 1, CTQ 3, and CTQ 5, it can be shown that:

$$\frac{d(HBr)}{dt} = \frac{2k_2k_3\left(\dfrac{k_1}{k_5}\right)^{1/2}(H_2)(Br_2)^{1/2}}{k_3 + k_4\dfrac{(HBr)}{(Br_2)}}$$

Compare the rate of reaction derived from the mechanism to the experimental rate law (1). Are the two rate laws the same?

8. In the initial stage of the reaction the rate is given by $2k_2 K_{eq}^{1/2}(H_2)(Br_2)^{1/2}$ where K_{eq} is the equilibrium constant that connects k_1 and k_5. Show how this relationship is obtained from CTQ 7.

9. From CTQ 8 it is possible to determine the rate constant k_2 and its temperature dependence experimentally. Suggest how this might be done.

Model 2: Some Thermodynamic Data for Step 2.

$$Br + H_2 \underset{k_4}{\overset{k_2}{\rightleftharpoons}} HBr + H$$

$$k_2 = A_2\, e^{\dfrac{-73.6\ \times\ 10^3\ J}{RT}}$$

Table 1.   Enthalpies of Formation

	$\Delta_f H^\circ_{298}$ (kJ/mol)
H(g)	218
H_2(g)	0
Br(g)	112
Br_2(g)	30.9
HBr(g)	−36.4

Critical Thinking Questions

10. What is the activation energy E_{a2}?

11. Using a reaction coordinate diagram and data from Table 1 find the activation energy, E_{a_4}, for

$$HBr + H \xrightarrow{k_4} H_2 + Br \ .$$

12. It is also possible to show experimentally that the ratio k_4/k_3 is independent of temperature. Find the activation energy E_{a_3}.

Information

The recombination of atoms requires essentially no activation energy. However, because a third body must be present to remove the energy of recombination, the rate of combination of atoms is slower than for that for polyatomic molecules. A three-body collision occurs about once for every 1000 two-body collisions. The rate of reaction of polyatomic free radicals is greater than that of atoms because polyatomic molecules may take up the energy of collision through internal rotations and vibrations.

Critical Thinking Questions

13. What is the activation energy E_{a_5}?

14. Calculate the activation energy E_{a_1}.

15. Summarize the steps of the mechanism corresponding to the activation energies.

Step	E_{a_i} (kJ/mol)
k_1 $Br_2 + M \longrightarrow 2Br + M$	
k_5	
k_2	
k_4	
k_3	

Exercises

1. Use the answers to CTQ 1, CTQ3 and CTQ5 to derive the rate law given in CTQ 7.

2. (a) Calculate the actvation energy for

$$H_2 + M \xrightarrow{k_6} 2H + M .$$

(b) This step produces intermediates that lead to subsequent reactions. In the proposed mechanism this step does not appear anywhere. The initiation step is considered to be

$$Br_2 + M \xrightarrow{k_1} 2Br + M .$$

Why is the step involving H_2 not considered to be the step that furnishes a radical species to initiate the reaction? Assume $T = 300$ °C.

(c) The termination step of the chain mechanism is considered to be

$$2Br + M \xrightarrow{k_5} Br_2 + M .$$

Why not $2H + M \xrightarrow{k_7} H_2 + M$?

3. Why does the step

$$HBr + Br \xrightarrow{k_8} H + Br_2 ,$$

which is the reverse of the step involving k_3, not appear in the mechanism?

4. Calculate the overall activation energy for the reaction.

5. Exercise 2 identifies initiation and termination steps of the reaction. Two of the steps of the mechanism given in Model 1 are called propagation steps. These are steps in which intermediate products react to produce further intermediates. Identify these two steps and explain your reasoning. One of the mechanistic steps is called inhibition. This results when a constituent is produced that slows the reaction. Identify this step and rationalize your choice.

Transition State Theory

Focus Question: What will be the effect of pressure on the rate of the reaction:

$$NO(g) + O_3(g) \rightleftharpoons \left[\begin{array}{c} O \\ O \end{array} \begin{array}{c} N \\ O \end{array} \begin{array}{c} \\ O \end{array} \right]^{\ddagger} \rightleftharpoons NO_2(g) + O_2(g)$$

Model 1: The Reaction Coordinate Diagram and the Rate Law for a Simple Reaction.

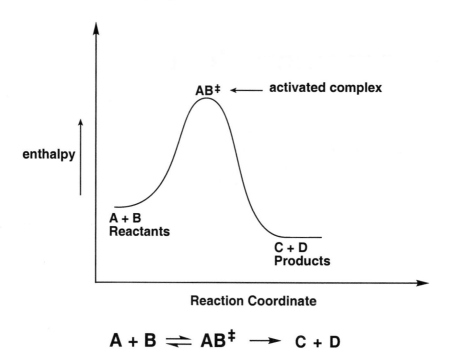

$$A + B \rightleftharpoons AB^\ddagger \longrightarrow C + D$$

The rate of a chemical reaction depends on the number of activated complexes passing over the potential barrier. The rate is therefore equal to the concentration of activated complexes times the frequency, f, with which the activated complex breaks apart to form products.

The rate law is given by

$$\frac{d(C)}{dt} = k_r(A)(B) \tag{1}$$

Critical Thinking Questions

1. Write an expression for the equilibrium constant, $K_c^{\ddagger}$ for the formation of the activated complex.

2. According to transition state theory what is the rate of the chemical reaction $\frac{d(C)}{dt}$?

Information

The rate at which the activated complex breaks apart is equal to the rate at which one of the vibrations of the activated complex leads to decomposition. The energy of that vibration is $E = hf$ where f is the frequency and h is Planck's constant. The classical energy of a vibration is kT. Therefore,

$$f = \frac{kT}{h} \qquad (2)$$

Critical Thinking Questions

3. Rewrite the rate law in CTQ 2 using equation (2)

4. Show that $\dfrac{d(C)}{dt} = K_c^{\ddagger} \dfrac{kT}{h} (A)(B)$.

5. In terms of transition state theory, what is the rate constant, k_r, for the reaction?

6. How is $\Delta G^{\circ \ddagger}$ related to $K_c^{\ddagger}$?

7. Verify that

$$k_r = \frac{kT}{h} K_c^{\ddagger} = \frac{kT}{h} e^{-\Delta G^{\circ \ddagger}/RT} = \frac{kT}{h} e^{-\Delta H^{\circ \ddagger}/RT} e^{\Delta S^{\circ \ddagger}/R}$$

8. It can be shown that the activation enthalpy, $\Delta H^{\circ\ddagger}$, is related to the Arrhenius' activation energy, E_a by $\Delta H^{\circ\ddagger} = E_a - RT$ (see Exercise 1).

If $k_r = Ae^{-E_a/RT}$ (A is not a function of temperature in the Arrhenius relationship.) show that

$$A = \frac{kT}{h} \, e \, e^{\Delta S^{\circ\ddagger}/R}$$

9. When the reaction is carried out at two different temperatures, show that

$$\ln \frac{k_{r2}}{k_{r1}} = \frac{-E_a}{R} \left(\frac{1}{T_2} - \frac{1}{T_1} \right)$$

Exercises

1. a) Give the derivative $\dfrac{d \ln K_c^{\ddagger}}{dT}$ in terms of $\Delta H^{\circ\ddagger}$. Show that $\ln k_r = \ln T + \ln K_c^{\ddagger} + \ln k/h$.

    b) From Arrhenius' theory

    $$k_r = Ae^{-E_a/RT}$$

    where E_a is the activation energy,

    show

    $$\frac{d\ln k_r}{dT} = \frac{E_a}{RT^2}$$

    and from transition state theory that

    c) $$\frac{d\ln k_r}{dT} = \frac{1}{T} + \frac{d\ln K_c^{\ddagger}}{dT}$$

    and hence that

    d) $$\frac{d\ln K_c^{\ddagger}}{dT} = \frac{E_a}{RT^2} - \frac{1}{T}$$

    Finally, show that

    e) $$\Delta H^{\circ\ddagger} = E_a - RT$$

2. The gas phase decomposition of a peroxide is first order and has rate constants $k_{100°C} = 4.3 \times 10^{-7}$ s^{-1} and $k_{200°C} = 2.9 \times 10^{-2}$s^{-1}. What are $\Delta G^{\circ\ddagger}$, $\Delta H^{\circ\ddagger}$, $\Delta S^{\circ\ddagger}$ at 100°C?

3. For the first order reaction

    $$CH_3NC(g) \rightleftharpoons CH_3CN(g)$$

    $$k_r = 4.0 \times 10^{13}\, e^{-\frac{160 \times 10^3}{RT}}$$

Model 2: The Entropy of Activation.

(1) $NO(g) + O_3(g) \rightleftharpoons \left[\begin{array}{c} \text{(structure)} \end{array} \right]^{\ddagger} \longrightarrow NO_2(g) + O_2(g)$

$\Delta S^{\circ \ddagger} = -40 \text{ JK}^{-1}\text{mol}^{-1}$

(2) $C\!-\!\!\!\!\stackrel{\displaystyle C}{\diagdown}\!\!C (g) \rightleftharpoons [C{=}C{-}C]^{\ddagger}$

$\Delta S^{\circ \ddagger} > 0$

Critical Thinking Questions

10. What does it mean when $\Delta S^{\circ \ddagger} \approx 0$?

11. Why is $\Delta S^{\circ \ddagger} < 0$ for reaction (1) above?

12. Why is $\Delta S^{\circ \ddagger} > 0$ for reaction (2) above?

13. In general what should be the sign of $\Delta S^{\circ \ddagger}$ for the decomposition of a ring compound? Explain.

14. How do the sign and magnitude of $\Delta S^{\circ \ddagger}$ affect the rate of a chemical reaction?

Effect of Pressure on Reaction Rates

Model 3: $k_r = K_c{}^{\ddagger}\,\dfrac{kT}{h}$.

Critical Thinking Questions

15. Show that

$$\left(\frac{\partial \ln k_r}{\partial P}\right)_T = \frac{-\Delta V^{\circ\ddagger}}{RT}$$

16. What is the meaning of the relationship given in CTQ 15?

Exercises

4. a) What is the effect of pressure at a fixed temperature on the rate of a reaction for which $\Delta V^{\circ\ddagger} = 0,\ \Delta V^{\circ\ddagger} > 0,\ \Delta V^{\circ\ddagger} < 0$?

   b) What does it mean when $\Delta V^{\circ\ddagger} = 0,\ \Delta V^{\circ\ddagger} > 0,\ \Delta V^{\circ\ddagger} < 0$?

   c) For the reaction

the rate goes up with increasing pressure. Speculate on the nature of the transition state.

5.   For the reaction

$$C_2H_5I(aq) + C_2H_5O^-(aq) \rightleftharpoons C_2H_5OC_2H_5(aq) + I^-(aq)$$

the rate constant increases slightly with increasing pressure. What might be the nature of the transition state?

6.   For the first order decomposition of C_2H_5I, the rate constant can be expressed as

$$k_r = 1.25 \times 10^{14}\, e^{\frac{-221,000}{RT}} \qquad E_a = \text{J mol}^{-1}$$
$$A = \text{s}^{-1}$$

$$C_2H_5I(g) \rightleftharpoons HI(g) + C_2H_4(g)$$

   a)   Find $\Delta G^{\circ\ddagger}$, $\Delta H^{\circ\ddagger}$, and $\Delta S^{\circ\ddagger}$ at 300 °C.

   b)   Comment on the value of $\Delta S^{\circ\ddagger}$.

   c)   How would pressure affect the rate constant of this reaction? Explain.

7.   For the decomposition of N_2O_5

$$N_2O_5(g) \rightleftharpoons NO_2(g) + NO_3(g)$$

$k_r = 1.72 \times 10^{-5}\ \text{s}^{-1}$ at 25°C and $k_r = 24.95 \times 10^{-5}\ \text{s}^{-1}$ at 45°C.

   a)   Calculate E_a for the reaction.

   b)   Find $\Delta S^{\circ\ddagger}$ and $\Delta H^{\circ\ddagger}$ at 25°C.

   c)   Suggest a possible transition state consistent with $\Delta S^{\circ\ddagger}$.

   d)   What would be the effect of pressure on the rate constant?

ChemActivity **M1**

Mathematics for Thermodynamics

Information

$$\lim_{\Delta x \to 0} \frac{\Delta y}{\Delta x} = \frac{dy}{dx}$$

$$y = f(x)$$

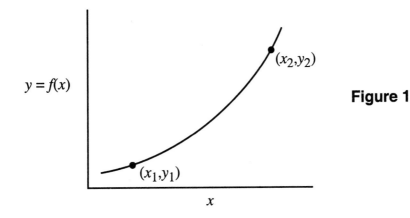

$y = f(x)$

Figure 1

Critical Thinking Questions

1. When a function is increasing what is the sign of $\frac{dy}{dx}$?

2. When a function is decreasing what is the sign of $\frac{dy}{dx}$?

3. When a function is neither increasing nor decreasing what is the value of dy/dx?

4.   In Figure 1 locate $\Delta y = y_2 - y_1$.

5.   In Figure 1 locate dy.

Exercises

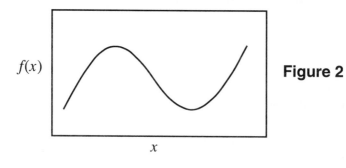

$f(x)$

x

Figure 2

1.   Roughly plot $\dfrac{df(x)}{dx}$ vs x where $f(x)$ is given in Figure 2.

2.   Roughly plot $\dfrac{d^2f(x)}{dx^2}$ vs x where $f(x)$ is given in Figure 2.

3. The plot in Exercise 1 shows two conditions for which $\frac{df(x)}{dx} = 0$, identify these conditions.

4. If the sign of $\frac{df(x)}{dx}$ changes from positive to negative upon passing through a point, what does that point represent?

5. If $\frac{df(x)}{dx}$ changes sign from negative to positive upon passing through a point, what significance does that point have?

6. If the sign of $\frac{d^2f(x)}{dx^2}$ is positive at a point, what is the significance of that point?

7. If the sign of $\frac{d^2f(x)}{dx^2}$ is negative at a point, what is the significance of that point?

8. What is the necessary condition that a function $f(x)$ have a point of inflection at a given point?

9. Find the first derivative of each of the following functions,

   $$y = x, \quad y = 1/x, \quad f(x) = \frac{Z(x)}{Y(x)}, \quad f(x) = Z(x)Y(x),$$

   where Z and Y are functions of x.

Information

In an integral the derivative of a function is given and it is required to find the function. The integral also gives the area under a curve plotted as $f(x)$ vs. x. Integration is the inverse of differentiation.

Critical Thinking Questions

6. If $F'(x) = \frac{df(x)}{dx}$

   find $\int F'(x)dx$.

7. If $F'(x) = \dfrac{df(x)}{dx}$

    find $\displaystyle\int_a^b F'(x)dx$

8. If $f(x)$ is known, can the area under the curve in CTQ 6 be determined?

9. If $f(x)$ is known, can the area under the curve in CTQ 7 be determined?

Information

$$d\ln x = \frac{dx}{x}$$

Critical Thinking Questions

10. Evaluate the integral

    $$\int f(x)dx$$

    if $f(x) = 1/x$.

11. Evaluate the integral

$$\int_a^b \frac{dx}{x}$$

12. Evaluate the integral $\int_{V_i}^{V_f} \frac{RT}{V} \, dV$ if R and T are constants and $V_i = 1$.

13. What is the area under the curve from $a = 1$ to $b = 2$ in CTQ 11?

14. Draw a rough graph of $1/x$ vs. x between $x = 0$ and $x = 5$. Then indicate the area described in CTQ 13.

15. Find $\displaystyle\int_{P_1}^{P_2} \frac{dP}{P}$.

Exercises

10. Evaluate the following integrals

$$\int_{x_1}^{x_2} d\ln x, \quad \int_{P_1}^{P_2} \frac{dP}{P^2}, \quad \int_{V_1}^{V_2} dV, \quad \int_{T_1}^{T_2} R dT \quad (R \text{ is the gas law constant.})$$

11. Show that $\displaystyle\int_{V_1}^{V_2} \frac{dV}{V} = \ln V_2/V_1$.

12. If $PV = RT$, is it possible to evaluate the integral $\displaystyle\int_{V_1}^{V_2} P dV$?

Model 1.

$$PV = nRT \quad R = 0.08206 \frac{\text{L-atm}}{\text{K-mol}}$$

Critical Thinking Questions

16. What variables does P depend upon in the model?

17. In general can you find the derivative $\dfrac{dP}{dT}$? State why or why not.

18. If n and V are constant find $\dfrac{dP}{dT}$ for Model 1.

19. If n and P are constant find $\dfrac{dV}{dT}$ for Model 1.

Model 2.

$$\left.\frac{\partial P}{\partial T}\right)_{n,V} = \frac{dP}{dT} \quad n,V \text{ constant}$$

Critical Thinking Questions

20. Find $\left.\dfrac{\partial V}{\partial T}\right)_{n,P}$, $\left.\dfrac{\partial P}{\partial V}\right)_{n,T}$, $\left.\dfrac{\partial V}{\partial P}\right)_{n,T}$ for Model 1.

Exercises

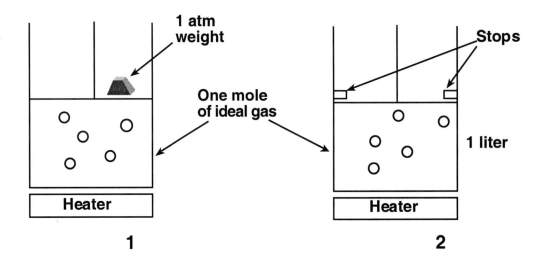

13. In 1 above how much will the volume change per degree of added heat?

14. In 2 above what happens if heat is supplied? Be quantitative.

15. Draw a diagram similar to 1 and 2 above for a process that is described by the derivative $\left.\dfrac{\partial T}{\partial V}\right)_{n,P}$.

16. Find the value of the derivative in Exercise 15.

Model 3.

$$P = P(T,V) \qquad\qquad \text{Eqn 1}$$

$$dP \;=\; \left.\frac{\partial P}{\partial T}\right)_{V} dT \;+\; \left.\frac{\partial P}{\partial V}\right)_{T} dV \qquad \text{Eqn 2}$$

Total   Partial   Partial
Differential Derivative Derivative

Critical Thinking Questions

21. State in words what Eqn 1 means.

22. State in words what Eqn 2 means.

23. If P = const, find $\dfrac{dV}{dT}$ in terms of other partial derivatives.

24. If V = const, find $\left(\dfrac{\partial V}{\partial T}\right)_P$ in terms of other partial derivatives.

25. If $PV = RT$ find $\left(\dfrac{\partial V}{\partial T}\right)_P$.

26. If $V = V(T,P)$ write the total derivative for V. If $P = $ const, what is dV?

Exercise

17. If $P = \dfrac{RT}{V} + \dfrac{(a+bT)}{V^2}$ R,a,b constants

     find $\left(\dfrac{\partial P}{\partial T}\right)_V$ and $\left(\dfrac{\partial V}{\partial T}\right)_P$.

Model 4.

If $F = F(x,y)$

$$dF(x,y) = X(x,y)dx + Y(x,y)dy$$

and if $\dfrac{\partial X}{\partial y} = \dfrac{\partial Y}{\partial x}$, the function is said to be exact.

Critical Thinking Questions

27. Is the function $F(x,y) = xy$ exact?

28. Is the function $F(x,y)$ exact if

$$dF(x,y) = ydx - xdy \; ?$$

29. Is $dF(T,P) = \dfrac{RT}{P}\, dP - R dT$ exact?

30. Is $dF(T,P) = \dfrac{R}{P}\, dT - \dfrac{RT}{P^2}\, dP$ exact?

Exercises

18. If $V = V(T,P)$ and $PV = RT$, show that

$$dV = \frac{R}{P}\, dT - \frac{RT}{P^2}\, dP.$$

Is dV an exact differential?

19. If $dw = -P dV$ and $PV = RT$, show that

$$dw = \frac{RT}{P}\, dP - R dT$$

Is dw an exact differential?

20.

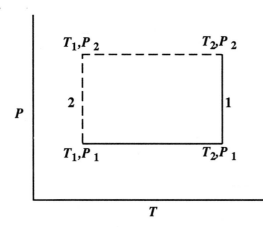

An ideal gas is at T_1, P_1. The two paths labeled 1 and 2 above may be used to change T_1, P_1 to T_2, P_2.

State in words how these two paths differ.

Find $\displaystyle\int_{T_1, P_1}^{T_2, P_2} dV$ along both paths.

Find $\displaystyle\int_{T_1, P_1}^{T_2, P_2} dw$ along both paths.

21. Use Exercises 18, 19 and 20 to describe the significance of an exact differential.

Model 5.

Taylor's Series

$$f(x) = f(0) + \frac{df(x)}{dx}\bigg|_{x=0} x + \frac{1}{2!} \frac{d^2f(x)}{dx^2}\bigg|_{x=0} x^2 + \frac{1}{3!} \frac{d^3f(x)}{dx^3}\bigg|_{x=0} x^3 + \ldots$$

Critical Thinking Questions

31. State in words what the above series expansion means.

32. Expand the function $f(x) = e^x$ in a Taylor's series.

33. If x is 0.001, represent e^x by a series accurate to 0.1%.

34. Of what use is a series expansion?

Appendix

TABLE A.1   Values of Selected Fundamental Constants

Speed of light in a vacuum (c)	$c = 2.99792458 \times 10^8$ m s^{-1}
Charge on an electron (q_e)	$q_e = 1.6021892 \times 10^{-19}$ C
Faraday's constant (F)	$F = 96{,}484.56$ C mol^{-1}
Planck's constant (h)	$h = 6.626176 \times 10^{-34}$ J s
Ideal gas constant (R)	$R = 0.0820568$ L atm mol^{-1} K^{-1}
	$R = 8.31441$ J mol^{-1} K^{-1}
Boltzmann's constant (k)	$k = 1.380662 \times 10^{-23}$ J K^{-1}
Avogadro's constant (N)	$N = 6.022045 \times 10^{23}$ mol^{-1}
Heat capacity of water	$C = 75.376$ J mol^{-1} K^{-1}

TABLE A.2   Selected Conversion Factors

Energy	1 J = 0.2390 cal = 10^7 erg
	1 cal = 4.184 J (by definition)
Temperature	K = °C + 273.15
	°C = (5/9)(°F − 32)
	°F = (9/5)(°C) + 32
Pressure	1 atm = 760 mm Hg = 760 torr = 101.325 kPa
	1 bar = 1×10^5 Pa
Mass	1 kg = 2.2046 lb
	1 lb = 453.59 g = 0.45359 kg
Volume	1 mL = 0.001 L = 1 cm^3 (by definition)
	1 L = 1 dm^3
Length	1 m = 39.370 in
	1 mi = 1.60934 km
	1 in = 2.54 cm (by definition)

TABLE A.3   SI Derived Units

Quantity	Name	Symbol	Dimension
Force	newton	N	kg m s^{-1}
Energy	joule	J	kg m^2 s^{-2}
Pressure	pascal	Pa	kg m^{-1} s^{-2}